UNDERSTANDING, EXPLANATION, AND SCIENTIFIC KNOWLEDGE

T0175613

From antiquity to the end of the twentieth century, philosophical discussions of understanding remained undeveloped, guided by a "received view" that takes understanding to be nothing more than knowledge of an explanation. More recently, however, this received view has been criticized, and bold new philosophical proposals about understanding have emerged in its place. In this book, Kareem Khalifa argues that the received view should be revised but not abandoned. In doing so, he clarifies and answers the most central questions in this burgeoning field of philosophical research: What kinds of cognitive abilities are involved in understanding? What is the relationship between the understanding that explanations provide and the understanding that experts have of broader subject matters? Can there be understanding without explanation? How can one understand something on the basis of falsehoods? Is understanding a species of knowledge? What is the value of understanding?

KAREEM KHALIFA is Associate Professor of Philosophy at Middlebury College. He has published a number of articles in philosophy of science, philosophy of social science, and epistemology.

UNDERSTANDING, EXPLANATION, AND SCIENTIFIC KNOWLEDGE

KAREEM KHALIFA

Middlebury College, Vermont

CAMBRIDGE
UNIVERSITY PRESS

University Printing House, Cambridge CB2 8BS, United Kingdom

One Liberty Plaza, 20th Floor, New York, NY 10006, USA

477 Williamstown Road, Port Melbourne, VIC 3207, Australia

4843/24, 2nd Floor, Ansari Road, Daryaganj, Delhi – 110002, India

79 Anson Road, #06–04/06, Singapore 079906

Cambridge University Press is part of the University of Cambridge.

It furthers the University's mission by disseminating knowledge in the pursuit of education, learning, and research at the highest international levels of excellence.

www.cambridge.org
Information on this title: www.cambridge.org/9781107195639
DOI: 10.1017/9781108164276

First published 2017

Printed in the United Kingdom by Clays, St Ives plc

A catalogue record for this publication is available from the British Library.

ISBN 978-1-107-19563-9 Hardback

For Lety, who understands me best.

Contents

List of Figures and Tables   *page* ix
Acknowledgments   x

1   The Philosophy of Understanding   1
   1.1   Kinds of Understanding   2
   1.2   Degrees of Understanding   3
   1.3   The Nexus Principle   6
   1.4   The Scientific Knowledge Principle   10
   1.5   The Explanation-Knowledge-Science (EKS) Model   14
   1.6   The Received View of Understanding   16
   1.7   A Look Ahead   20

2   An Illustration: Bjorken Scaling   23
   2.1   Preliminaries and Clarifications   23
   2.2   Bjorken on Scaling   27
   2.3   A Lack of Understanding   34
   2.4   Comparison with Batterman   42
   2.5   Comparison with De Regt   43
   2.6   Conclusion   50

3   Understanding and Ability   51
   3.1   The Classic Ability Argument   52
   3.2   Resisting the Classic Ability Argument   54
   3.3   The Updated Ability Argument   63
   3.4   Pritchard on Cognitive Achievements   65
   3.5   Hills on Cognitive Control   70
   3.6   Grimm on Grasping   72
   3.7   Conclusion   79

4   Objectual Understanding   80
   4.1   The Lay of the Land   82
   4.2   Robust Breadth   92
   4.3   Austere Breadth   100
   4.4   Robust Coherence   104

4.5 Austere Coherence   III
4.6 Conclusion   123

5 Understanding Without Explanation?   125
   5.1 Lipton's Framework   126
   5.2 Putting Explanation Back into Understanding   129
   5.3 Examples of Modal Understanding   131
   5.4 The Galileo Example   135
   5.5 Unification via Tacit Analogy   139
   5.6 Tacit Understanding of Causes   144
   5.7 Conclusion: A Rapprochement?   150

6 Understanding and True Belief   154
   6.1 Getting Our Bearings   155
   6.2 Historical Arguments   158
   6.3 Idealization Arguments   166
   6.4 Conclusion   181

7 Lucky Understanding   183
   7.1 The Luck Questions   184
   7.2 Passé Compatibilism   186
   7.3 Nouveau Incompatibilism   194
   7.4 Argumentum ex Scientia   201
   7.5 Conclusion   207
   7.6 Excursus: Gettier Luck   209

8 The Value of Understanding   212
   8.1 The Basic Value Question   213
   8.2 The Distinctive Value Question   221
   8.3 Conclusion: Understanding, Philosophy,
      and Scientific Knowledge   234

Bibliography   236
Index   246

Figures and Tables

Figures

1.1 to 1.4 Different Explanatory Structures   *page* 9
      5.1 Tacit Representation of Causal Knowledge   148
      7.1 Conditions for Interventions   202

Tables

      1.1 Different Kinds of Non-Explanatory Understanding   2
      7.1 Differences Between the Barn Façade, Strict Nero, and   192
           Pritchard's Nero Examples

Acknowledgments

Many people deserve credit for their support as I developed the ideas for this book. I apologize in advance for any omissions, particularly of audience members where this material was presented. Of course, all errors, deficiencies, and problems in this book are my own.

The book's earliest ideas were developed during my first stay at the University of Pittsburgh's Center for Philosophy of Science (2010–2011), and the final touches were made there during my most recent stay (2016). It is hard to imagine a better place at which to philosophize about science. I wish to thank John Norton, Jim Lennox, and Edouard Machery for directing the Center, as well as the various Center Fellows who provided feedback on parts of this book: Yann Benetreau-Dupin, Pierluigi Barrotta, Matt Brown, Anjan Chakravartty, José Díez, Heather Douglas, Alison Fernandes, Tobias Henschen, Hylarie Kochiras, Bert Leuridan, P. D. Magnus, Daniele Muttini, Elisabeth Nemeth, Viorel Pâslaru, Laura Perini, Richard Samuels, Sam Schindler, J. D. Trout, Pete Vickers, Ioannis Votsis, and John Worrall. Additionally, I would like to thank several faculty members at the University of Pittsburgh and Carnegie Mellon for their personal and intellectual hospitality during my stints at the Center: Bob Batterman, Jim Bogen, Mazviita Chirimuuta, David Danks, Kevin Kelly, Peter Machamer, Ken Manders, Sandy Mitchell, Joel Smith, Jim Woodward, Porter Williams, and Kevin Zollman.

I would also like to thank the community of philosophers working on understanding, whose feedback has been invaluable: Christoph Baumberger, Georg Brun, Henk de Regt, Kate Elgin, Victor Gijsbers, John Greco, Stephen Grimm, Chris Kelp, Soazig Le Bihan, Mark Newman, Michael Strevens, Emily Sullivan, and Daniel Wilkenfeld. I cannot imagine the many foolish things that would have ended up in this book in the absence of their feedback and intellectual camaraderie. Henk, Kate, and Stephen deserve special thanks for their thoughtful feedback on my early work on understanding; Mark, for pushing me to clarify crucial aspects of my account; Emily and Daniel, for reading complete versions of an earlier draft of this book.

Hilary Gaskin, Sophie Taylor, and other hard-working people at Cambridge University Press deserve special mention for their professionalism, efficiency, and thoroughness. I would also like to thank two anonymous referees for their feedback on an earlier draft of this manuscript.

Portions of several articles are found throughout Chapters 2, 4, 5, and 7. While I have revised these ideas substantially and spun them into a more systematic account, I thank the publishers for their permission in allowing me to reprint the following:

- Khalifa, Kareem (2012), "Inaugurating Understanding or Repackaging Explanation?" *Philosophy of Science* 79 (1):15–37. With permission of University of Chicago Press.
- Khalifa, Kareem, and Michael C. Gadomski (2013), "Understanding as Explanatory Knowledge: The Case of Bjorken Scaling," *Studies in History and Philosophy of Science Part A* 44 (3):384–392. With permission of Elsevier.
- Khalifa, Kareem (2013), "Is Understanding Explanatory or Objectual?" *Synthese* 190 (6):1153–1171. With permission of Springer.
- Khalifa, Kareem (2013), "The Role of Explanation in Understanding," *British Journal for the Philosophy of Science* 64 (1):161–187. By permission of Oxford University Press.
- Khalifa, Kareem (2013), "Understanding, Grasping, and Luck," *Episteme* 10 (1):1–17. Reprinted with permission of Cambridge University Press.

I would also like to thank Middlebury College for its support. I am especially grateful for the College's generous Undergraduate Collaborative Research Fund that availed me of excellent research assistants over several summers: Gabe Doble, Yannick Doyle, Mike Gadomski, Daniel Ramirez, and Conor Simons. Special thanks to my colleague, Noah Graham, for discussions about statistical mechanics that helped me work through certain issues in Chapter 6.

While this project contains only a modicum of material from my dissertation, I owe my two graduate school mentors—Mark Risjord and Bob McCauley—an intellectual and professional debt that I cannot possibly repay. Let me briefly mention two. First, in his *Woodcutters and Witchcraft*, Mark cashed out the understanding characteristic of the social sciences using an epistemology of explanation. While our epistemologies of explanation differ, this book nevertheless pushes a similar line for the natural sciences. Similarly, at key junctures in this book, I make analytic epistemology subservient to scientific practice. This is a stance that I learned from Bob—even if, in assuming our shared stance, I add my own idiosyncratic flourishes.

As many of my friends and acquaintances know, my two dogs, Frida and Cleo, are essential pillars in my support network. (They are sleeping peacefully beside me as I write this after a long day.) The humans in that network are also notable. I am forever indebted to my parents, Ezzat and Shadia Khalifa, for their love and for constantly supporting my intellectual pursuits throughout my childhood. Finally, my wife, Leticia Arroyo Abad, is my anchor, my inspiration, and the best thing that has ever happened to me. This book is dedicated to her.

CHAPTER I

The Philosophy of Understanding

We seek to understand a great many things – the heights of the heavens; the depths of the oceans; motions and emotions; flora and fauna; friends, family, and foes; minds and machines; markets and morals; and much more. But understanding is only faintly understood. When do we limn the deep structures of the natural and social world? When do our attempts at grasping these structures misfire?

Understanding can be understood in different ways. An *empirical* inquiry might use experiments, interviews, surveys, and the like to uncover the mechanisms that make understanding possible, or it might track the historical arc of our various conceptions of understanding. By contrast, I will be pursuing a *theoretical* inquiry, in which I use various formal and conceptual apparatuses to construct a model of understanding.

As a *philosophical* theorist of understanding, my instruments are the tools of my specific trade. Thus, while the interesting theoretical concepts found in the cognitive and social sciences undoubtedly illuminate various aspects of understanding, my preferred instruments come from epistemology and the philosophy of science. Recent philosophical work on understanding straddles these venerable fields – crisscrossing with larger questions about justification, knowledge, and other cognitive achievements; the point and purpose of scientific explanation; the role of models and idealizations in scientific theories; and the pragmatic aspects of scientific inquiry. These and other philosophical forays provide a repository of ways to plumb the hidden depths of understanding.

Ideally, lively interdisciplinary crosstalk at both the empirical and theoretical levels should be achieved. Nevertheless, given the nascent state of the field, I hope that I will be forgiven for attempting to slay only a small handful of understanding's many dragons. And, lest I seem too indifferent to other areas of study, I will frequently argue that attention to scientific practice is an effective cure for the overzealousness that infects philosophy's more speculative organs.

1.1. Kinds of Understanding

Even among philosophical accounts of understanding, there are many kinds that can be studied and scrutinized. I will chiefly be interested in *explanatory* understanding (i.e., the understanding characteristic of good explanations). Paradigmatically, such understanding can be expressed as a kind of understanding-why; for instance, when we say, "Susan understands why the sky is blue."

For ease of locution, I will treat "explanatory understanding" and "understanding-why" as synonyms. Having said this, explanatory understanding can be expressed without a "why." For instance, I take my account to cover nearby examples, such as "Susan understands what causes the sky to be blue" and "Susan understands how it is that the sky is blue."

Additionally, I will be restricting myself to understanding of *empirical* phenomena. I make no claims about understanding in ethics, aesthetics, mathematics, or logic, for example. This assumes, of course, that these are not empirical domains. My hankerings for a panoptic empiricism notwithstanding, I will stake no claim on these tangled conceptual thickets.

While other kinds of understanding, as listed in Table 1.1, are also worthy of study, they will not be my focus:

Table 1.1: *Different Kinds of Non-Explanatory Understanding*

Kind of understanding	Typical Complement	Examples
Propositional	*that* + declarative sentence	I understand that you might not enjoy reading this book.
Broad Linguistic	The name of a language	Schatzi understands German.
Narrow Linguistic	*what* + a linguistic expression + *means*	Schatzi understands what "Ich bin Berliner" means.
Procedural	*how* + infinitive	Miles understands how to play trumpet.
Non-explanatory Interrogative	Embedded question that does not seek an explanation as its answer (most who, where, what, and when questions)	I understand who my friends are. I understand where my friends will be going. I understand what my friends are doing. I understand when my friends need a good laugh.

The list in Table 1.1 is meant to be neither exhaustive nor particularly precise. Furthermore, some of these kinds of understanding are undoubtedly necessary (though not sufficient) for certain instances of explanatory understanding. Nevertheless, the devils in their details need not be exorcised here – my account of explanatory understanding should be compatible with whatever turns out to be the best accounts of these kinds of non-explanatory understanding.

My fixation on explanatory understanding includes one notable detour. Chapter 4 discusses explanatory understanding's relationship to something that philosophers call *objectual* understanding. Roughly stated, it is the understanding one has of a subject matter. It takes as its complement a noun phrase (e.g., "Niels understands quantum mechanics"). I will argue that explanatory understanding already captures anything philosophically important about objectual understanding. Hence, aside from considerations of linguistic convenience, we have no real need for the latter.

1.2. Degrees of Understanding

However, even when we focus on explanatory understanding of empirical phenomena, our philosophizing can only begin after a further clarification. Such understanding admits of *degrees*. For instance, recall our protagonist, Susan, who understands why the sky is blue. But now assume that she is a leading atmospheric physicist. Presumably, her understanding of the sky's blueness would be quite robust, involving a grasp of many causal factors, connections with deep theoretical principles, experimental results, methodologies, and so on. By contrast, we might credit Susan's freshman student, Bill, with understanding why the sky is blue even though he grasps only a tiny fraction of the information at Susan's disposal. In short, Susan's understanding is *better* than Bill's.

How do we navigate these different degrees of understanding? We might analyze a kind of *minimal* understanding by identifying the conditions that are necessary for any understanding whatsoever. Alternatively, we might analyze a *maximal* or *ideal* kind of understanding, which would be a mirror image of minimal understanding. In either case, a way of *comparing* different people's understanding would be a handsome prize, as we could then describe the full spectrum.

I will start with minimal and comparative principles of understanding, and then derive other gradations accordingly. The details of these principles needn't concern us quite yet. For now, I simply want to sketch how we could cover the gamut of understanding from my preferred starting point.

To begin, what does it mean to understand *better*? This is where comparative principles enter our happy scene. They will have the following form:

> *Schema for Comparative Understanding: Ceteris paribus, S_1* understands why *p* better than S_2 if and only if S_1 has minimal understanding of why *p* + *X*.

Obviously, this is just a sketch; we still need to fill in the value of *X*. Nevertheless, even these bare bones invite a few *obiter dicta*.

First, note that one can only have *better* understanding if one has *minimal* understanding. For instance, Susan could not understand why the sky is blue better than Bill if Susan does not have at least *some* understanding of why the sky is blue. This point seems obvious enough that I will leave it implicit hereafter.

Second, *X* may have several moving parts. Thus, some comparisons will be messy, as different individuals may have greater understanding with respect to one dimension of *X*, while being further away with respect to others. For instance, one person may have a tenuous grasp of the evidence that confirms her extremely accurate explanation; another may have a firm grasp on the evidence that supports her less accurate explanation. In this kind of situation, the *ceteris paribus* clause is violated, and there may be no clear way to determine who has better understanding. Alternatively, the relative importance of these differences may be specified by the context in which the comparison is made. Of course, in other cases, one understander strictly dominates another, and the comparison is relatively straightforward. This is presumably the case with Susan and Bill.

We can then use these comparative principles to derive a conception of ideal understanding:[1]

> *Ideal Understanding: S* ideally understands why *p* if and only if it is impossible for anyone to understand why *p* better than *S*.

Note that here, "impossible" means "logically impossible" and not merely "humanly possible." This is an ideal, so we might as well aim high!

From this ideal, we can then pair a minimal account of understanding with a contextualist semantics to make sense of non-comparative or "outright" understanding, which earns its keep in the ample space between the minimum and the ideal:

---

[1] I've learned much about how to think about degrees of understanding from Kelp (2015). One small difference: Kelp takes maximum/ideal understanding as his starting point and gleans comparative and outright conceptions of understanding from there. These are only methodological points. I briefly cover the substantive differences between our views in Chapter 4.

Outright Understanding: "*S* understands why *p*" is true in context *C* if and only if *S* has minimal understanding and *S* approximates ideal understanding of why *p* closely enough in *C*.

Thus, on this view, contexts dictate how closely one must approximate the ideal. For instance, my understanding of why my car moves consists of little more than my facility in depressing the gas pedal. This understanding has proven serviceable on the boulevards of Vermont, but has not served me nearly as well in the arena of car repair. Thus, given the rules of the road, I understand why my car moves, but, given the standards in the shop, I do not. Thankfully, my local mechanic ably meets the standards in these latter contexts.

Furthermore, an account of outright understanding presupposes a prior theoretical account of *both* minimal *and* better understanding (the latter by way of ideal understanding.) If a context's standards of approximating the ideal fall below the threshold of minimal understanding, then they are too weak to do their job. However, if outright understanding did not appeal to principles of better understanding, then it would be identical with minimal understanding.

Finally, I introduce another concept of understanding mostly for convenience. Sometimes, we need not specify how much understanding agents actually have. Suppose that a person knows her way around a car engine, but we do not know how deeply that understanding runs. In such cases, it will be useful to use "generic understanding" attributions:

Generic Understanding: *S* has some understanding of why *p* if and only if "*S* understands why *p*" is true in some context *C*.

In other words, we can think of generic understanding as having understanding to some degree or another.

As we will see, much of the extant philosophical literature on understanding frames its questions as if understanding were an all-or-nothing affair. Thus, the key questions have been, "Does understanding require explanations? Must it consist of mostly true information? Is it a species of knowledge?" I will be recasting many of these questions so as to capture the fact that understanding admits of degrees.

To summarize, I will take accounts of objectual, ideal, outright, and generic understanding to spring from the fountainhead of minimal and comparative accounts of explanatory understanding. Let's begin to build this fountainhead, beginning with two comparative principles.

1.3. The Nexus Principle

A natural suggestion is that explanatory understanding is the possession or "grasp" of an explanation. For instance, to understand why the sky is blue is to have a correct explanation of why the sky is blue. However, multiple factors contribute to the azure above. For instance, blue light's wavelength is relatively short, it is scattered in all directions by molecules in Earth's atmosphere, the spectrum of light emission from the sun does not distribute all frequencies in equal proportion, the high atmosphere absorbs violet light, individuals' positions to the sun changes at different times of the day – and don't get me started on our eyeballs' intimate workings.

Presumably, one's understanding of the sky's blueness increases as one gathers more of these correct explanatory factors and also as one learns how these correct factors hang together. Let the *explanatory nexus* of p be the set of correct explanations of p as well as the relations between those explanations.[2] I suggest the following as the first of my two comparative principles of understanding:

> The Nexus Principle: Ceteris paribus, if S_1 grasps p's explanatory nexus more completely than S_2, then S_1 understands why p better than S_2.[3]

This raises three important questions. First, what is a correct explanation? Second, which relations between explanations furnish understanding? Third, what makes one person's grasp more "complete" than another's? I address each in turn.

1.3.1. A "Theory" of Explanation

I face an interesting challenge in depicting the nexus' main inhabitants: correct explanations. On the one hand, because I would like the larger points about understanding to swing freely of any of my idiosyncrasies about explanation, this favors being relatively noncommittal about the nature of explanation. On the other hand, if I am too noncommittal, then my claims about understanding become inscrutable. In writing this book, the following seems to have struck the right balance:

---

[2] In conversation, it appears that some people individuate explanations slightly differently than I do. If you find yourself stumbling on this, then simply replace the count-noun "explanations" with the mass-noun "explanatory information." Thus, the nexus of p is the totality of explanatory information about p, and grasping more of this information improves one's understanding.

[3] Recall from Section 1.2 that I am leaving implicit the requirement that S_1 has better understanding than S_2 only if S_1 has minimal understanding. Parallel points apply throughout my discussion of better understanding.

q (correctly) explains why *p* if and only if:

(1) *p* is (approximately) true;
(2) *q* makes a difference to *p*;
(3) *q* satisfies your ontological requirements (so long as they are reasonable); and
(4) *q* satisfies the appropriate local constraints.

Hereafter, "explains" is elliptical for "correctly explains," unless otherwise noted. Furthermore, I will follow the time-honored philosophical pretension of using the Latin *explanandum* to denote *p* (the statement to be explained) and *explanans* to denote *q* (the statement which does the explaining).

Let's discuss each condition in turn. The first condition is relatively uncontroversial. For instance, nothing correctly explains why the sky is paisley, why Mitt Romney won the 2012 presidential election, or why electrons are positively charged.

Similarly, many theorists of explanation agree upon the second condition. Consider the claim that blue light's short wavelength explains why the sky is blue. This entails that blue light's wavelength *makes a difference* to the sky's color. A common way of unpacking this is in terms of counterfactual dependence: had blue light's wavelength been longer, then the sky would have been a different color (Lewis 1986; Woodward 2003). For the most part, I will assume this counterfactual approach to difference-making. However, my arguments should not be affected if an alternative approach to difference-making (e.g., Strevens 2008) were countenanced in its stead.

The third condition, that the explanans satisfy your (presumably reasonable) ontological requirements, is designed to elide complicated issues concerning scientific realism. For the purposes of this book, realists will hold that the explanans *q* should be treated in the same manner as the explanandum – it should be (approximately) true. By contrast, many antirealists deny that our best explanations have true explanantia. A prominent antirealist alternative only requires them to be empirically adequate (van Fraassen 1980). Empirical adequacy is, roughly speaking, the requirement that a theory says only true things about directly observable entities, processes, and the like. This applies not only to those phenomena that are actually observed, but also to those that are *observable* – including all past, present, and future phenomena. By contrast, anything a theory says about *un*observable entities – paradigmatic examples of which are subatomic particles, the curvature of spacetime, species, mental states, and social structures – may be false without forfeiting explanatory correctness.[4]

Many philosophers would chafe at this proposal and require that both the explanans and the explanandum are approximately true. I will be walking a fine line here: none of my arguments obliges someone to abandon this stronger commitment if he or she is so inclined. However, neither do my arguments compel someone to adopt these stronger commitments if he or she is disinclined. Feel free to swap out this third condition with a requirement that satiates your realist longings without suspicion or shame. For all practical purposes, this will only bear on Chapter 6.

Finally, we get to the fourth and most cryptic of my requirements on explanations – that they satisfy "local constraints." I take the first three conditions on explanation to be "global" constraints: they apply to explanations anywhere we find them. But, gaze deeply into my soul, and you will see a card-carrying explanatory pluralist staring back at you: the relevance of many explanatory features depends on the specific explanandum, the standards of the discipline, and the interests of the inquirer. To get a taste of these local constraints, in subsequent chapters we will see that only some of our explanations:

- represent causal structure;
- deploy asymptotic reasoning;
- represent mechanisms;
- represent non-causal, contrastive, probabilistic relations
- unify phenomena into a single framework;
- use idealizations; and
- represent potential interventions.

I emphasize that these local constraints must be satisfied *in addition* to the three global constraints I place on explanation. Hence, I take the global constraints to assuage concerns that I'm being too slippery, and the local constraints to afford me enough flexibility to remain faithful to the diversity of scientific practice and to grant certain assumptions to my interlocutors that will enliven the dialectic by avoiding foot-stamping impasses about what is – and is not – an explanation.

---

[4] The labels "scientific realist" and "scientific antirealist" aren't the tidiest tags to use here. For instance, there are entity realists (Cartwright 1983; Hacking 1982) and structural realists (Worrall 1989) who reject the idea that our best explanations are approximately true. These folks might well count as antirealists in this book.

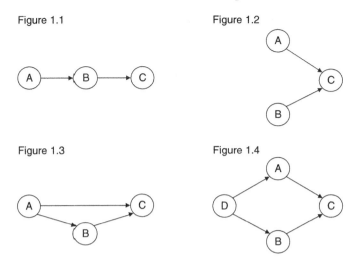

Figures 1.1 to 1.4: Different Explanatory Structures
A letter at the tail of an arrow does the explaining (explanans); a letter at the head of an arrow is explained (explanandum).

1.3.2.   Inter-Explanatory Relationships

The explanatory nexus includes not only correct explanations, but also the *relationships* between them. What is characteristic of these relationships? One obvious relationship is that of *relative goodness*. Some explanations are *better* than others, even if both are correct. For instance, the presence of oxygen is explanatorily relevant to any fire's occurrence. However, very rarely will the presence of oxygen be the *best* explanation of a fire, in part because oxygen is also frequently present in the absence of fires. Per the Nexus Principle, grasping these sorts of facts enhances one's understanding.

However, superiority and inferiority are not the only relationships between correct explanations. For instance, consider the four scenarios in Figure 1.1. to 1.4. Suppose that in all of these cases, both A and B are correct explanations of C. However, even if, e.g., A is a better explanation of C than B, this would not say anything about the *structure* that these diagrams represent.[5] Quite clearly a person who could not distinguish these different

---

[5] Perhaps indirect explanations of C, such as A in Figure 1.1 or D in Figure 1.4, should not be part of the nexus. I submit that if this is so, then they will figure in one's scientific knowledge of the nexus. As sketched in Section 1.4 and discussed more extensively in subsequent chapters, this is also part of my account of understanding. Consequently, grasping indirect explanations will figure somewhere in my account of understanding; exactly where is not terribly important.

explanatory structures would not understand C as well as someone who did. For instance, a person who knew that A only explains C through B in Figure 1.1, or that A and B are independent of each other in Figure 1.2, or that D is a common explanation or "deep determinant" of both A and B in Figure 1.4, and so forth. Intuitively, this person has a better understanding of C than a person who did not grasp these relationships. Undoubtedly, explanations can stand in other relationships that figure in the nexus.

1.3.3.   Completeness of Grasp

The Nexus Principle pins the goodness of one's understanding to the *completeness* of one's grasp of explanatorily relevant information. But what does the completeness entail? I submit that a person's completeness of grasp is proportional to each of the following:

- The number of correct explanations and inter-explanatory relations grasped
- The quality/importance of the explanations and inter-explanatory relations grasped
- The level of detail of the explanations and inter-explanatory relations grasped

It would be a mistake to think of my account of completeness in quantitative terms – e.g., we do not typically count the number of explanations, nor will anything so detailed will be required. For the most part, we will encounter situations in which there is a stock of explanatory information that two or more inquirers both grasp and then some further bit of explanatory information that is unique to one. Indeed, science frequently progresses along just these lines. For the purposes of this book, these sorts of comparisons require no quantitative metric of one's grasp of the nexus.

1.4.   The Scientific Knowledge Principle

Essentially, I have taken a truism – that explanations are answers to explanation-seeking questions – as the springboard for the Nexus Principle. However, understanding involves more than this. The Nexus Principle appeals to the capacity of an agent to "grasp" explanatory information. However, to analyze understanding in terms of grasping is to swap out an enigma for a mystery.

Let me betray some philosophical biases. Since understanding is ubiquitous, I would suspect that grasping is ubiquitous, too. So, in my estimate, grasping should be something quite mundane. As a result, I'm not out to give a philosophical account of "grasping" that requires an extravagant phenomenology or that populates our psychology with exotic mental states. I unpack this black box with a second comparative principle:

> *The Scientific Knowledge Principle: Ceteris paribus*, if S_1's grasp of p's explanatory nexus bears greater resemblance to scientific knowledge than S_2's, then S_1 understands why p better than S_2.

In this case, I'll say that S_1's understanding is more *scientific* than S_2's. Note that this further explains why an atmospheric physicist's understanding of the sky's blueness is greater than her student's. In addition to possessing more explanatory information, the way in which that information is possessed involves mastery of more evidence, background theories, and methods – the physicist not only has more of the nexus in her grasp, but that grasp is also firmer. Indeed, understanding appears to be intimately connected to expertise (Wilkenfeld, Plunkett, and Lombrozo 2016). Since I focus on explanations of empirical phenomena, and scientists are the leading experts on this front, the Scientific Knowledge Principle has some initial plausibility.[6]

This raises some preliminary questions. What is scientific knowledge of an explanation? How should we understand a grasp's resemblance to that knowledge?

1.4.1.   Scientific Knowledge

The Scientific Knowledge Principle sums up my account of grasping: it's nothing more than a cognitive state bearing some resemblance to scientific knowledge of some part of the explanatory nexus. I defend this thesis in Chapter 3. Before then, let me clarify a few things about the Scientific Knowledge Principle.

Much like explanation, my goal is to say just enough about scientific knowledge without holding my account of understanding hostage to any epistemological idiosyncrasies, while at the same time saying enough to address the puzzles in subsequent chapters. To that end, I offer the following:

---

[6] Of course, this does raise larger issues about how we reconcile the "manifest and scientific images," as Sellars (1963) famously framed the issue. I briefly address this issue in Section 1.5.

S has scientific knowledge that q *explains why* p if and only if the safety of S's belief that q *explains why* p is because of her scientific explanatory evaluation.

A person's belief is safe just in case her belief-forming process could not easily have led to a false belief (Pritchard 2005; Sainsbury 1997; Sosa 1999; Williamson 2000). Requiring understanders to have safe beliefs fills out various boxes on any epistemologist's checklist. For instance, standard epistemologies require knowledge to include true belief plus some anti-Gettier condition, both of which fall out of the requirement that a belief is safe. I further discuss understanding's relationship to true belief in Chapter 6 and its relationship to safety in Chapter 7.

What then of the remaining component of scientific knowledge – what I call *scientific explanatory evaluation*? I will assume that scientific evaluation of an explanation (SEEing) has three invariant features: consideration, comparison, and belief-formation.[7]

First, scientists typically *consider* many of the plausible potential explanations of the phenomenon of interest. More precisely, scientists create models of the nexus, since they also consider relationships between different explanatory factors.[8] Sometimes, consideration requires generating new hypotheses from scratch, or (more commonly) it only involves countenancing explanations that have been generated by others (e.g., by doing a careful lit review). Regarding the first stage, a potential explanation is a correct explanation modulo the third condition listed in Section 1.3.1 – which concerns the explanans' ontological credentials (e.g., truth/empirical adequacy). However, some potential explanations are implausible (e.g., explaining Newton's death by appeal to alien laser guns). These need not be considered. While I offer no precise account of plausibility, typical considerations include the explanation's fit with accepted background theories and simplicity. I should also stress that some plausible explanations are incorrect. For realists, such explanations will be ones that fail to be approximately true; for antirealists, such explanations fail to satisfy a more modest requirement (e.g., empirical adequacy).

Let's turn to the second feature of SEEing. Scientific explanatory evaluation typically involves the ability to *compare* the potential explanations that have been considered. Here, scientists use the best methods, evidence, and perhaps other non-evidential considerations characteristic of the natural and

---

[7] My view draws some inspiration from Lipton's (2004, 61) account of Inference to the Best Explanation (IBE). I take no stand on IBE's cogency.

[8] Hereafter, I'll use the language of "plausible potential explanations" as shorthand for the more barbarous "plausible potential models of the explanatory nexus."

social sciences as we currently find them. These considerations clearly favor some explanations over others. In paradigmatic cases of comparison, one explanation is the "winner" of these comparisons, though sometimes multiple explanations are good along different dimensions, and often these explanations are complements rather than competitors. Comparisons tease out these sorts of relationships.

In the last stage of SEEing, scientists *form (doxastic) attitudes* based on the comparisons just discussed. Scientists believe that clear winners in the prior stage of comparison are correct, believe that clear losers incorrect, and assign appropriate degrees of belief about the middle of the pack. In principle, some non-doxastic attitudes, such as acceptance, might also figure in this stage, but I'll put this to the side until Chapter 6.

Finally, returning to the definition of scientific knowledge, the safety of one's explanatory commitments must be *because* of one's SEEing. As virtue epistemologists hold that knowledge consists of true beliefs that are because of an agent's abilities (Greco 2009; Sosa 1991, 2007; Zagzebski 1996), and SEEing is an exercise of many abilities that (ideally) results in true beliefs, this aspect of my account is naturally glossed in virtue-epistemological terms. I return to this in Chapter 8.

1.4.2.   Resemblance

As with the Nexus Principle's notion of completeness, the Scientific Knowledge Principle appeals to *resemblance* in order to capture the comparative account of understanding that I'm chasing. What is meant by "resemblance to scientific knowledge?" The preceding account of scientific knowledge suggests that this resemblance will be proportional to:

- The number of plausible potential explanations that the agent has considered
- The number of considered explanations that the agent has compared using scientifically acceptable methods and evidence
- The scientific status of the methods and evidence that the agent used to compare the explanations
- The safety of the agent's beliefs about explanations
- The accuracy of the agent's beliefs about explanations

As far as I can tell, this account of resemblance to scientific knowledge suffices for achieving most of the book's argumentative objectives. Additionally, for reasons discussed in Chapter 6, I add the following dimension of resemblance to scientific knowledge:

- The variety of ways that the agent can use explanatory information so as to achieve different scientific goals

Furthermore, since I take better understanding to be a fundamental unit of analysis in discussions of understanding, it will often be useful to describe the spectrum of cognitive states up to and including scientific knowledge. For ease of reference, I call these states "epistemic statuses." For instance, in contexts with very low standards for understanding – such as the examples involving Bill's understanding of the sky's blueness or my understanding of my car's movement – perhaps one only needs to have an approximately true belief in a correct explanation, even though one's belief is unsafe or one lacks scientific evidence. Thus, talk of grasping can always be replaced by a more specific epistemic status (e.g., approximately true beliefs, non-scientific knowledge, scientific knowledge). In other words, we can always swap out the placeholder – the buzzword "grasping" – with something more pedestrian and informative.

1.5. The Explanation-Knowledge-Science (EKS) Model

Earlier, I claimed that we need accounts of both minimal and better understanding to furnish a comprehensive treatment of understanding. I will be arguing that the Nexus and Scientific Knowledge Principles are all that we need to provide a satisfactory philosophical account of better understanding, i.e.:

(EKS1) S_1 understands why p better than S_2 if and only if:
   (A) *Ceteris paribus*, S_1 grasps p's explanatory nexus more completely than S_2; or
   (B) *Ceteris paribus*, S_1's grasp of p's explanatory nexus bears greater resemblance to scientific knowledge than S_2's.[9]

What, then, of minimal understanding? In Chapter 3, I argue for the following:

(EKS2) S has minimal understanding of why p if and only if, for some q, S believes that q *explains why* p, and q *explains why* p is approximately true.

Since these principles rely on explanation, knowledge, and science as their key concepts, I call these claims the *EKS Model of Understanding*. (If it

---

[9] Recall that I am leaving it implicit that S_1 has minimal understanding of why p; see footnote 3.

floats your boat, pronounce it as if it were the X-Model, or, if you prefer to reminisce about former fashionistas, the Ex-Model.)

Principles (A) and (B) – the Nexus and Scientific Knowledge Principles – may occasionally offer conflicting counsel, namely when one person grasps more explanatory detail in a less scientific way than another, or vice versa. I stipulate that this is precisely when the *ceteris paribus* clause is violated. Nevertheless, we can settle these conflicts in a few ways. First, it may be that in the particular context of comparison, complete understanding is prized more dearly than scientific understanding (or vice versa). Alternatively, we can advert to more specific evaluative vocabulary to capture the differences between the two agents. For instance, we might say that one person's understanding is more complete and another's is more scientific.

Given the discussions in Section 1.2, I also take EKS1 and EKS2 as sufficient for providing adequate accounts of ideal, outright, and generic understanding, and I will generally speak of these derivative concepts as part of the EKS Model. Very roughly, ideal understanding is maximally scientific knowledge of a complete explanatory nexus. Different contexts will indicate how complete and scientific one's (outright) understanding must be. When one meets those context-specific requirements and also has minimal understanding, one understands in that context. When there is at least one context in which one adequately approximates scientific knowledge of some part of the nexus, one has some (i.e., generic) understanding.

This notion of outright understanding allows me to place scientific understanding and "everyday" understanding on a single continuum. An example of the latter would be the understanding of why a window is shattered when one learns of the antics of a rock-throwing vandal. It requires no special scientific knowledge – or, as I would prefer to think of it, it requires only scant resemblance to scientific knowledge. Because understanding admits of degrees, the spectrum of understanding looks roughly like this:

> Minimal understanding < Everyday understanding < Typical scientist's understanding < Ideal understanding

I claim that the differences between minimal and everyday understanding (when they exist) amount to different context-specific standards about how closely one must approximate ideal understanding. Parallel points apply to the differences between everyday understanding and a typical scientist's understanding.

1.6.   The Received View of Understanding

I am, by training, a philosopher of science interested in explanation, and my view of understanding shows my debt to my intellectual ancestors. For much of the twentieth century, no concept of understanding-why garnered significant philosophical attention. Nevertheless, something like a "received view" emerged from several oft-repeated, offhand remarks made by various philosophers of science interested in analyzing explanation. I see the EKS Model as a more regimented descendant of the received view. The remainder of this book aims to defend my view from criticisms of the received view. To that end, I want first to get a clearer sense of what the received view is, and then to identify its extant challenges.

First, the received view holds that understanding is a species of knowledge. Peter Lipton (2004, 30) states it pithily: "Understanding is not some sort of super-knowledge, but simply more knowledge: knowledge of causes." However, Lipton was certainly not the first to trumpet this idea. In his masterful "Four Decades of Scientific Explanation," Wesley Salmon summarizes the state of play among philosophers of science working on explanation in the late 1970s as follows:

> explanations enhance our understanding of the world. Our understanding is increased (1) when we obtain knowledge of the hidden mechanisms, causal or other, that produce the phenomena we seek to explain, (2) when our knowledge of the world is so organized that we can comprehend what we know under a smaller number of assumptions than previously, and (3) when we supply missing bits of descriptive knowledge that answer why-questions and remove us from particular sorts of intellectual sorts of intellectual predicaments. (Salmon 1989, 134–135)[10]

Additionally, as these quotes indicate, both Lipton and Salmon take understanding to be knowledge of an explanation. Lipton advocates for a causal model of explanation. Hence, when he describes understanding as knowledge of causes, he's treating understanding as explanatory knowledge. The three items in Salmon's list correspond to the way he carves up the philosophical landscape in the 1970s: ontic/causal, inferential, and pragmatic models of explanation.

Indeed, even those who didn't play the knowledge game were willing to put their chips on the table when it came to understanding's link to explanation. For example, although Hempel (1965, 337) does not state that understanding is a kind of knowledge, he nevertheless asserts:

---

[10] See Grimm (2006) for further quotes from more philosophers of science to this effect.

the argument shows that, given the particular circumstances and the laws in question, the occurrence of the phenomenon was to be expected; and it is in this sense that the explanation enables us to *understand* why the phenomenon occurred.

Of course, Hempel equated explanation with deductive-nomological (DN) arguments, so this says little more than that understanding tracks with explanation. But such a stance toward understanding is not limited to DN accounts of explanation. Speaking only of his own view, Salmon, a leading advocate of the causal-mechanical model of explanation, writes:

> causal processes, causal interactions, and causal laws provide the mechanisms by which the world works; to *understand* why certain things happen, we need to see how they are produced by these mechanisms. (Salmon 1984, 132; emphasis added)

Similarly, Kitcher asserts:

> Science advances our *understanding* of nature by showing us how to derive descriptions of many phenomena, using the same patterns of derivation again and again, and, in demonstrating this, it teaches us how to reduce the number of types of facts we have to accept as ultimate (or brute). (Kitcher 1989, 432)

Once again, talk of using as few patterns of derivation to derive as many phenomena as possible is the core idea underlying Kitcher's unificationist account of explanation.

Importantly, the received view was not proffered as a considered or systematic account of understanding. A quick review of the literature suggests that earlier philosophers of science "neglected" understanding as an area of research for two reasons. First, many of them thought that understanding was *irrelevant* because it is a merely psychological or pragmatic phenomenon of no epistemic significance.[11] Second, even if understanding was not merely psychological afterglow, it was nevertheless thought to be *redundant*, being replaceable by explanatory concepts without loss.[12]

Despite understanding's peripheral nature in the philosophy of science through the end of the twentieth century,[13] the previous quotes suggest the following:

---

[11] While this view dates at least as far back as Hempel (1965), its contemporary guise comes in Trout's (2002, 2005, 2007) charges that the sense of understanding is a highly unreliable cognitive faculty. For critiques of Trout's view, see De Regt (2004, 2009a, 2009b), De Regt and Dieks (2005), and Grimm (2009b).

[12] For an opposing view, see Wilkenfeld (2014).

[13] Indeed, the received view might antedate the twentieth century by two millennia. Greco (2013) suggests that Aristotle endorsed something quite similar to it.

> *The received view: S* understands why *p* if and only if there exists some *q* such
> that *S* knows that *q explains why p*.[14]

For better or for worse, this is how current philosophers of understanding construe the received view. As should be clear, the received view and EKS Model are closely related, as both put explanation and knowledge at the helm. Moreover, given that the main proponents of the received view have been philosophers of science, by "knowledge" they might well have meant "*scientific* knowledge."

Given these affinities, it might seem as if I'm putting a large target on my back, as the received view has been the target of lively philosophical scrutiny in the twenty-first century. One prominent objection to the received view is that understanding involves "special abilities" that explanatory knowledge does not require. For instance, some have claimed that we can know an explanation by way of passive testimony or rote memory without achieving understanding (De Regt 2009b; Hills 2009; Kvanvig 2003; Newman 2012, 2014; Pritchard 2008, 2009a, 2010). Such examples put pressure on the received view's contention that explanatory knowledge is sufficient for understanding. Invariably, this has been driven by the idea that understanding requires abilities in a way that knowledge does not. Call this the *Classic Ability Question:*

• Do understanding and knowledge require the same abilities?

(I'll explain why I've added the "Classic" tag to this and other questions later.)

Other authors have resisted the received view's privileging of explanatory information. This objection takes two forms. First, some claim that objectual understanding differs from explanatory understanding. They claim that an explanation only provides deep understanding insofar as it figures in the more comprehensive understanding of a subject matter typically associated with objectual understanding (Carter and Gordon 2014; Elgin 2007; Kvanvig 2003) or that we can gain objectual understanding from classifications, discoveries, and unexplainable systems even when we lack explanatory information (Gijsbers 2013, 2014; Kelp 2015; Kvanvig 2009b). Call this the *Classic Objectual Question:*

• Is objectual understanding the same as explanatory understanding?

---

[14] Often the right-hand side of this biconditional is stated as "*S* knows why *p*" or "*S* knows that *p because q*." I treat these as synonymous to my preferred gloss of the received view.

Second, others claim that understanding can be achieved through non-explanatory means (Kvanvig 2009b; Lipton 2009). To choose but one example, a person may understand how a device works by way of her ability to correctly manipulate that device while lacking explanatory resources. Call this the *Explanation Question:*

• Does understanding require explanation?

Additionally, some friends of understanding have challenged the received view's privileging of *true* information. If epistemologists agree on nothing else, they agree that knowledge is factive (i.e., if one knows that *p*, then *p* is true). However, some have alleged that understanding can depart from the truth (De Regt 2015; Elgin 2004, 2007, 2009b; Riggs 2009). For instance, a useful fiction may provide greater understanding than its more accurate counterpart. Call this the *Truth Question:*

• Does understanding require true belief?

Clearly, a negative answer to the Truth Question threatens the idea that understanding is a species of knowledge.

Additionally, much recent work grants that understanding trades in true, explanatory information but denies that it is a species of knowledge. These claims focus on knowledge's fourth, "anti-Gettier" condition (Kvanvig 2003, 2009b; Pritchard 2008, 2009a, 2010, 2014). Those partial to this line argue that while knowledge cannot be the product of certain kinds of luck, understanding can. Call this the *Classic Luck Question:*

• Is understanding incompatible with epistemic luck?

The preceding questions concern the *nature* of understanding. However, epistemologists will have noticed a sizeable omission from my list – what I call the *General Value Question:*

• Do understanding and knowledge have the same kinds of epistemic value?

Quite clearly, proponents of the received view will downplay the differences between knowledge and understanding and will therefore be inclined to answer the General Value Question affirmatively. Yet one of the driving forces behind the recent philosophical interest in understanding is a worry that knowledge lacks a certain kind of value that understanding possesses (Kvanvig 2003, 2009b; Pritchard 2008, 2009a, 2010, 2014).

Note that these questions are not only interesting in their own right, but also motivate the recent interest in understanding. If all of them can be answered affirmatively, the received view's guiding idea– that understand-ing should be an afterthought – seems more or less correct: the philoso-phical study of understanding is redundant given the more battle-tested inquiries into explanation and knowledge.

1.7.   A Look Ahead

In what follows, I motivate and defend my account of understanding from the challenges raised against the received view. In one sense, my responses to these challenges lead precisely to the position just can-vassed: give me an epistemology of scientific explanation, and I will give you everything you could have wanted from a philosophy of understanding.

However, given my emphasis on degrees of understanding, the Classic Ability, Classic Objectual, and Classic Luck Questions must be recast when pitched at the EKS Model. For instance, the Classic Ability Question cuts too coarsely. Testimony often affords someone a modest amount of understanding, but that understanding would improve with more significant exercises of ability. This suggests the following:

> *Updated Ability Question:* Which abilities, if any, improve our understanding?

The EKS Model requires that these understanding-improving abilities play a role in scientific knowledge. As I show in Chapter 3, alternative accounts of the abilities that figure in understanding are defensible only insofar as they respect this constraint.

Similarly, some proponents of objectual understanding – whom I dub "robust objectualists" – query the sufficiency of the received view. As with the Classic Ability Question, we should "update" our ways of thinking, viz.

> *Updated Objectual Question:* Are comparably demanding cases of objectual and explanatory understanding identical in all philosophically important ways?

In Chapter 4, I argue that robust objectualists' arguments are defused if we think of robust objectual understanding as the equivalent of having lots of explanatory understanding. For instance, a sentence such as "Niels under-stands quantum mechanics" is just shorthand for "For many *p* about quantum mechanics, Niels understands why *p*."

Much like the Classic Ability and Classic Objectual Questions, the Classic Luck Question, covered in Chapter 7, should be replaced by:

> *Updated Luck Question:* Does understanding improve as it becomes less lucky?

As with abilities, I hold that less demanding instances of understanding tolerate luck, while also insisting that the diminution of luck through scientific means is a mark of greater understanding.

The Explanation and Truth Questions interrogate the received view's claim to have provided necessary conditions on understanding. These questions remain unchanged and are addressed in Chapters 5 and 6, respectively. However, when targeted at the EKS Model, they are rightly aimed at my account of minimal understanding (EKS2). By showing that even minimal understanding requires approximately true explanations, it follows that more demanding instances of understanding also require approximately true explanations.

Finally, as we'll see in Chapter 8, the General Value Question requires some recasting, but in different ways than we'll see with the remaining, "Classic Questions." In particular, improvements in understanding become the chief bearers of epistemic value, but in ways that are inextricably linked to their resemblance to scientific knowledge.

Earlier, I noted that my forebears tended toward indifference with respect to understanding, seeing it as either irrelevant or redundant. My relationship to this stance is wonderfully conflicted. On the one hand, since explanatory knowledge isn't merely some noetic buzz, I do not think that understanding is irrelevant. On the other hand, since we have reasonably good accounts of knowledge and explanation that can do most of the work required of understanding, deep departures from these accounts are redundant.

There are still other ways in which the deflationary stances of my intellectual heroes do not overlap with my own. Figures such as Hempel and Salmon sought to provide *heavyweight* theories of explanation: their accounts trade exclusively in global constraints. As already discussed, mine is of a more *lightweight* variety: I am advancing an account of explanation that seeks to be as noncommittal as possible while achieving the book's larger objectives. My account of explanation is thin because of its steady diet of local constraints. In effect, where heavyweight theories of explanation would replace those local constraints with some substantive and universal condition on explanation, I mostly have to shrug and say, "Look at the relevant science." So, in this regard, I am more deflationary than traditional purveyors of the received view.

On the other hand, the old guard said very little about the *epistemological* components of understanding. I, on the other hand, have offered an analysis of scientific knowledge of explanation. In this regard, I am perhaps less deflationary than they are: I believe that a substantive epistemology underlies understanding.

Nevertheless, it might still be regarded as deflationary in two respects. First, more than a few philosophers of understanding aim to show that bread-and-butter analyses of knowledge cannot capture understanding's core features. This book's arguments reveal this aim to be misplaced. This is essentially to claim, as partisans of the received view did, that fancy new theories of understanding are redundant.

Second, while the epistemological components of my view combine various elements of antiluck epistemology and virtue reliabilism, there is still another point in my analysis where I must again shrug and simply point to the science. In particular, the stage of SEEing in which explanations are compared hinges on using the *best available scientific methods and evidence*. Certain experimental tests are apt for adjudicating between certain explanations, theoretical considerations may figure more prominently in adjudicating between others, non-experimental empirical tests may figure more prominently in still other adjudications, and so forth. In other words, I am suspicious of general criteria by which to compare explanations. Compared to epistemologies that take a heavyweight stance to such questions (Harman 1986; Lipton 2004; Lycan 1988; Thagard 1978, 1992), this stance may well look deflationary.

Thus, in cases where I engage in lightweight theorizing, scientific practice *dictates* certain parameters of the EKS model. However, other parts of my view *account* for certain aspects of scientific practice. In effect, this sharpens the old dichotomy of "irrelevant or redundant." Philosophizing about *some* aspects of understanding is irrelevant – not because those aspects are merely psychological, but because discovering what they are is best answered by appealing to local facts about scientists' methodologies, so little in the way of philosophy is useful. On the other hand, philosophizing so that *other* aspects of understanding *seem radically distinctive* is misguided – because an epistemology of explanation, suitably articulated, already accounts for the relevant aspects of scientific and everyday practice and thereby renders these overreaching conjectures redundant.

At any rate, that's enough lowbrow meta-philosophy. Let me practice what I preach and delve into some highbrow physics to illustrate how the EKS model applies to flesh-and-blood scientists.

CHAPTER 2

An Illustration: Bjorken Scaling

Talking with both epistemologists and philosophers of science is one of the best things about working on understanding. In the larger philosophical profession, the lack of engagement between these two sub-disciplines is striking. Despite science providing some of our most celebrated cognitive achievements, most epistemologists disregard the work of philosophers of science. Despite epistemologists' highly sophisticated and general analyses of knowledge, philosophers of science are more likely to talk with scientists than with epistemologists. Working on understanding gives me some hope that these trends might change.

Having said that, talking with both epistemologists and philosophers of science is also one of the *worst* things about working on understanding. Given the splintering of these sub-disciplines, it's rather maddening to figure out whether I should be trying to convince others by way of simple, elegant, but highly artificial thought experiments, or detailed, messy, but highly realistic episodes from the history of science. Working on understanding gives me little hope that these sorts of problems will be resolved in a way that will leave either side terribly happy.

For now, let's take our model for a spin in the messy world of science. In particular, let's look at an episode in the history of science where we have a clear upturn in understanding and see if we can make sense of it using the Explanation-Knowledge-Science (EKS) Model. If it works when talking about high-energy physics experiments, I'm cautiously optimistic that it will work when we turn to high-energy thought experiments.

2.1. Preliminaries and Clarifications

In the preceding chapter, I proposed the following:

(EKS1) S_I understands why p better than S_2 if and only if:

(A)   *Ceteris paribus,* S_1 grasps p's explanatory nexus more completely than S_2; *or*

(B)   *Ceteris paribus,* S_1's grasp of p's explanatory nexus bears greater resemblance to scientific knowledge than S_2's.

In accord with the preceding chapter, I will call (A) the "Nexus Principle" and (B) the "Scientific Knowledge Principle."[1]

Let me unpack just enough of these principles to make the philosophical points about the physics that I will discuss. As I noted in Chapter 1, the EKS Model is a descendent of the received view. As Grimm (2006) notes, the received view has an illustrious fan club: Achinstein, Kitcher, Lipton, Salmon, and Woodward endorse the view that understanding is a species of explanatory knowledge. However, these authors focus on the concept of explanation while paying relatively little attention to the concept of knowledge. By contrast, the EKS Model says little about explanation, but says more about the epistemology of understanding.

My choice for focusing less on the concept of explanation is deliberate. One of the earlier generation's big mistakes was an unhealthy attachment to a single model of explanation. By contrast, I accept a plurality of explanatory relations (Díez, Khalifa, and Leuridan 2013; Khalifa, Millson, and Risjord forthcoming).[2] Thus, I do not subscribe to an exclusively causal model, an exclusively unificationist model, or the like. Instead, these – and others – are appropriate in different contexts As we'll see later, this allows me to countenance explanations that only inhabit fairly exotic locales, such as asymptotic explanations in certain branches of theoretical physics.

Before proceeding, two additional points are in order. Recall that the explanatory nexus of p is the set of correct explanations and their relations. My first point is that I use the phrase "correct explanation" to sidestep certain debates about scientific realism.[3] Thus, even if antirealists turn out to have the more defensible view of explanation, I would still hold that understanding improves as we grasp more correct explanations and their relationships; I would only add that the concept of *correctly explains* should be glossed antirealistically.

For the purposes of illustration, consider two scenarios. In the first, everyone's favorite antirealist, Bas van Fraassen (1980), turns out triumphant

---

[1]   I also discussed my account of minimal understanding (EKS2) in the previous chapter. It will become fairly obvious that this threshold has been met in this example, so I omit its discussion for ease of prose.

[2]   See Sections 1.3.1 and 2.3.1 for further details.

[3]   See Section 1.3.1 for a more detailed discussion of why I hold that correct explanations have explanantia that are "at least" empirically adequate.

in the final analysis about explanation. In the second, the most granite-jawed realist gets the better of van Fraassen. In either case, the EKS Model would hold that understanding improves as one grasps more true statements of the form *q correctly explains p*. But here's the twist: if van Fraassen is right, we EKS'ers would follow him in holding that *q* correctly explains *p* if and only if (crudely glossed) *q* and *p* are part of an empirically adequate theory, and *q* stands in the contextually appropriate relevance relation to *p*. By contrast, if the realist wins the day, then we'd dutifully follow her in asserting that *q* correctly explains *p* if and only if (again, crudely glossed) *q* and *p* are approximately true and stand in the appropriate objective dependency relationship. (Having said this, I can't stress enough that I'm *not* taking a stance on the realism issue and that Chapter 6 provides the relevant details.)

Second, let's unpack what we mean by "scientific knowledge of an explanation." As discussed in Chapter 1, I define this as follows:

> *S* has scientific knowledge that *q explains why p* if and only if the safety of *S*'s belief that *q explains why p* is because of her s̲cientific e̲xplanatory e̲valuation (SEEing).

Recall that SEEing has three features: consideration, comparison, and belief formation. First, scientists typically can *consider* many of the plausible potential explanations of the phenomenon of interest. Sometimes consideration requires generating new hypotheses from scratch, or (more commonly) it only involves countenancing explanations that have been generated by others.

Let's turn to the second feature of SEEing. Scientific explanatory knowledge typically involves the ability to *compare* the potential explanations that have been considered. Here, scientists cite relevant evidence (and perhaps other, non-evidential scientific factors) that favors some explanations over others. Insofar as simplicity, scope, mechanism, conservatism, analogy, testability, unification, fruitfulness, and other theoretical virtues figure in scientific practice, they do so at this stage. In paradigmatic cases of comparison, one explanation is the "winner" of these comparisons, though sometimes multiple explanations are good along different dimensions, and often these explanations are complementary rather than competitors. Comparisons tease out these sorts of relationships.

Finally, scientists *form doxastic attitudes* based on the comparisons just discussed. Scientists believe that clear winners in the prior stage of comparison are correct, believe that clear losers are incorrect, and assign appropriate degrees of belief about the middle of the pack. Earlier, I mentioned that I take scientific knowledge to require true belief that was not fortuitous (i.e., that could not easily have been false). In other words, scientific knowledge requires safe belief. Following Pritchard (2009b, 34), I define safety thus:

> S's belief is safe *iff* in most near-by possible worlds in which S continues to form her belief about the target proposition in the same way as in the actual world, and in all very close near-by possible worlds in which S continues to form her belief about the target proposition in the same way as the actual world, her belief continues to be true.

My description of scientific explanatory evaluation clarifies how S "continues to form her belief about the target proposition in the same way as the actual world." In the present context, this means that, in the relevant possible worlds, S continues to believe that *q explains p* by:

- Considering the same class of potential explanations of p as she did in the actual world, and
- Comparing those explanations in the same way and on the basis of the same evidence as she does in the actual world.

Additionally, scientific knowledge of *why p* typically presupposes scientific knowledge *that p*, though what this latter knowledge entails depends mightily on the phenomenon or explanandum of interest. For instance, we'll look at examples in which our explananda are known by way of running scattering experiments in particle accelerators. This won't always be a viable method for knowing explananda in other branches of physics (e.g., astronomy), much less in chemistry and the special sciences (though it brings a smile to my face to imagine cultural anthropologists having a go at the large hadron collider). In other words, the value of p goes a long way toward telling us what counts as scientific knowledge that p.

While I present SEEing as necessary for scientific knowledge of an explanation, I would not be too upset if this turned out to be false. In this case, SEEing would be one member of a larger team of scientific methods and processes for acquiring scientific knowledge. As someone with broadly naturalistic leanings, I believe that scientific practice should set the final roster and the starting lineup.

I'll say more about scientific knowledge and SEEing in subsequent chapters (especially Chapters 3 and 7). In the remainder of this chapter, I'll illustrate the EKS Model's plausibility by examining the history of Bjorken scaling. In the late 1960s, James "BJ" Bjorken made a novel prediction about a certain kind of scaling. Bjorken used rather abstract theoretical tools that were largely opaque to a majority of physicists – even the experimental physicists who performed the experiment that confirmed

Bjorken's prediction. Shortly thereafter, Richard Feynman explained scaling in terms of "partons," giving Bjorken's complex mathematical model and the scaling phenomenon a physical interpretation that was intelligible to a much wider range of physicists. Feynman's parton model allowed experimental physicists to run further experiments that ruled out other potential explanations of scaling throughout the early 1970s. I'll tell this tale in further detail, illustrating how the EKS Model captures some fairly central features of understanding.

I've chosen this example for several reasons. First, as an enterprise, physics provides exemplary understanding of the empirical world. Second, in this particular case, a period of confusion (i.e., lack of understanding) arose and was subsequently resolved. Studying this transition illustrates how the EKS Model accounts for the conversion from a well-confirmed but opaque phenomenon into an object of understanding. I'll conclude my discussion by highlighting several advantages that the EKS Model has over Batterman and De Regt's accounts of understanding. In the process, I anticipate some objections to my EKS-influenced narrative.

2.2.   Bjorken on Scaling

In this section, I first describe Bjorken's prediction (Section 2.2.1). Then, I show that, even before Feynman's interpretation, the EKS Model's requirement of known explananda is satisfied (Section 2.2.2), and I discuss Bjorken's explanation of the phenomenon he predicted (Section 2.2.3). At the beginning of the next section, I'll highlight the aspects whereby Bjorken's explanation failed to provide understanding.

2.2.1.   *Early History of Bjorken Scaling*

In the latter half of 1967, a team of researchers from the Stanford Linear Accelerator (SLAC) group and the Massachusetts Institute of Technology (MIT) measured the "scatter" that results from firing a beam of electrons at a proton target. More precisely, they measured the *cross-section* σ, the likelihood of an interaction between particles.

These experiments were designed to discover basic properties of subatomic particles. The *hadron–lepton* distinction is especially important in classifying these particles. Hadrons are subatomic particles that are affected by nuclear or "strong" forces, whereas leptons are immune to such forces. Thus, neutrons and protons (as well as, e.g., kaons and pions) are hadrons, and electrons (as well as muons and neutrinos) are leptons.

Physicists represent scattering experiments as BT → X, where B refers to the beam particle, T to the target particle, and X to the particles that result from their interaction. The most important SLAC-MIT experiments of this time can be represented as ep → X, where e refers to electrons and p to protons. The research team examined two kinds of scattering phenomena. The first, *elastic* scattering, involves interactions in which beam and target particles – in this case, electrons and protons, respectively – retain their identities. Thus, elastic scattering experiments can be represented as ep → ep. The second, *inelastic* scattering, involves interactions in which the proton need not retain its identity (i.e., ep → eX, for all X). The team did not attempt to identify the various particles comprising X in these inelastic scattering experiments.

Prior to the SLAC-MIT experiments, it was assumed that cross-sections for both elastic and inelastic scattering would fall off sharply when electron beams were fired at higher energy levels and scatter was measured at larger angles. Quantum electrodynamics (QED), the dominant theory of the time, assumed that electrons interacted as hard, point-like entities,[4] while protons had a diffuse, soft structure extended over a finite volume of space. If QED were correct, then there would be very little scatter at high energies and large angles, as soft protons would only permit electrons to strike glancing blows. Elastic scattering experiments performed prior to 1967 were consonant with this result, as cross-sections for electron–proton scattering were much smaller at larger angles than cross-sections for electron–electron scattering.

The SLAC-MIT team fired the electron beam at higher energies than their predecessors. All of their results were consistent with prior theory and experiment, *except* for the surprising discovery that the cross-sections for electron–proton and electron–electron interactions are roughly the *same* as *inelastic* scattering at high energies and large angles – what is called the "deep inelastic" region. In other words, the electron–proton interactions have much higher cross-sections than was previously expected for deep inelastic scattering. *Contra* QED, this suggests that the proton is composed of hard point-like entities.

Bjorken, working at the SLAC theory group, was one of the few physicists unsurprised by this result because, in 1966, he had already predicted it using a then-esoteric mathematical framework in quantum field theory called current algebra.[5] More precisely, Bjorken predicted that

---

[4] Strictly speaking, QED only assumes this in first-order approximation. I've omitted this detail in what follows.
[5] First published in Bjorken (1967).

the absolute energy of an experiment does not determine the cross-section of electron–proton scattering, which is consistent with the SLAC-MIT team's surprising result that these cross-sections do not decrease at higher energy levels.

According to Bjorken, the cross-section of deep inelastic scattering – hereafter σ_{DIS} – is determined instead by the ratio of the energy loss of the scattering electrons v to the momentum transfer between the electron and the proton q. This point is intimately related to Bjorken scaling, which I'll discuss later. Bjorken suspected that a more direct consequence of scaling could be gleaned from the experimenters' results, so, in April 1968, he urged them to plot two well-known functions that represent the proton's structure (W_1 and W_2) against what would later be known as the *Bjorken scaling variable*[6] $\omega = -q^2/Mv$. Here, M is the proton's mass. As Bjorken predicted, the results fell on unique curves – now called *Bjorken scaling curves*.

2.2.2.   *Knowledge of Explananda*

Earlier, I suggested that scientific knowledge of *why* something is the case (an *explanandum*) typically requires knowledge *that* it is the case. After discussing which explananda are in play in this example, I'll discuss some of the ways in which the experimenters came to know these explananda.

There are two principal explananda in this little slice of particle physics. The first explanandum is the scattering phenomenon, i.e.:

Why is $\sigma_{DIS}/\sigma_{MOTT} \approx 1$ (rather than <1)?

Here, σ_{MOTT} is the cross-section of electron–electron scattering, and the parenthetical contrast indicates the result predicted by QED. Since it was known that electrons were hard, but QED assumed that protons are soft, the experiments suggest that electron–proton scattering more closely resembles electron–electron scattering than QED predicts.

A second explanandum, concerning scaling, requires a bit more unpacking. At the most general level, a scaling law (Bjorken's or otherwise) is a function f such that $f(cx) \propto f(x)$, where c is a constant. Thus, changing the size or scale of the function's argument preserves the shape of the function. A simple example of scaling is the equation that expresses a square's area A as a function of the square of the length l of one of its sides (i.e., $A(l) = l^2$). Regardless of the length of l – i.e., regardless of the scale of the square – this relationship holds. In this case, we say that area *scales like* length squared.

---

[6] This is Bjorken's original notation. Subsequently, the scaling variable was defined as $x = q^2/2Mv$.

In Bjorken scaling, the principle is the same, but the relationships and quantities are a bit more involved. Specifically, the Bjorken scaling laws are:

$$W_1 = F_1(\omega); \textit{and}$$
$$\nu W_2 = F_2(\omega).$$

Just as area scales like length squared, W_1 scales like $F_1(\omega)$, and W_2 scales like $F_2(\omega)$. These scaling laws serve as our second explanandum. In short, inquirers were interested in understanding why these scaling relations hold (e.g., why an increase in ν necessitates a proportional increase in F_2).

Other explananda figure in subsequent parts of this story. In a later section, I will argue that these are best seen as part of the scientists' evaluation of the parton explanation. For now, I turn to whether the scientists had *knowledge* of the following explananda:

$$\sigma_{DIS}/\sigma_{MOTT} \approx 1 \text{ (rather than } < 1); \qquad (E1)$$

$$W_1 = F_1(\omega) \text{ and } \nu W_2 = F_2(\omega). \qquad (E2)$$

Minimally, it would appear that the scientists believed that these two phenomena were not mere artifacts. Moreover, by the lights of our best current science, these beliefs are true and were delivered by reliable methods. Thus, quite plausibly, the scientists knew these two explananda.

For instance, one might think that the surprising results about deep inelastic scattering (E1) would have provided sufficient reason for members of the SLAC-MIT team to reject QED's assumption about a soft proton. However, members of the team debated whether the large number of electrons that had lost their energy in the deep inelastic scattering experiments was the result of their colliding with the hard, point-like constituents of the proton or the result of the electrons radiating photons during the collision (a well-known phenomenon). In order to rule out the latter possibility, members of the SLAC team used a computer-driven, time-intensive process to make "radiative corrections" in order to disambiguate the data. In the spring of 1968 – nearly a *year* after the experiment was performed! – the radiative corrections were complete, revealing that the high number of low-energy electrons could not be explained away by photon radiation. Since the same data were used to plot the scaling curves (E2), parallel points apply to this explanandum. These precautions are precisely the kinds of things that distinguish knowledge from a fortuitously true belief.

2.2.3.   *Bjorken's Explanation*

Of course, simply *knowing that* (E1) and (E2) are true does not amount to *understanding why* they are true. The Nexus Principle suggests that understanding advances when we start to grasp correct explanations of these two phenomena. To that end, Bjorken's account of scaling is a near-complete sketch of an *asymptotic explanation*, which has received its most lucid philosophical exposition from Robert Batterman (2002).

To explain universal phenomena asymptotically involves identifying and eliminating details that are irrelevant to the behavior in question. To this end, one employs sophisticated mathematical techniques to examine the asymptotic behavior of the appropriate governing functions. Give or take a few niceties, the general schema for (one kind of) asymptotic explanation is:

Asymptotic Explanation Schema[7]
Explanation Target
Why does the same **pattern of behavior** emerge in diverse physical **systems**?
Explanatory Pattern
The **pattern of behavior** can be expressed as a mathematical **function**.
Various **details** about these **systems** are constant in the **asymptotic limit** of this **function**.
Given the **underlying microphysics** of the **system**, differences in these **details** are irrelevant to the **pattern of behavior**.

Here, the boldface phrases are variables that are filled in according to the explanandum of interest. I will fill in this schema later.

Certain variables in this explanation schema deserve further clarification. By **details**, I simply mean parameters that are used to describe the system. For instance, in the example discussed later, these details include beam energy and scattering angles of electrons. By **underlying microphysics**, I mean the interactions on the microscopic level of the various details of the systems.

---

[7] I borrow the use of explanation schemas (though not asymptotic ones) from Thagard (1999). In broad outline, this schema replicates Batterman's (2002, 44) three criteria of asymptotic explanations:

- "... the explanation involves some kind of asymptotic analysis.
- The universality is the result of the stability under perturbation of the underlying microscopic details [i.e., the behavior of a system tends to remain the same even when basic features of the system are changed].
- The stability under perturbation (or 'structural stability') is explained by the underlying microscopic physics of the systems being investigated."

The technical variable in this explanation schema, the **asymptotic limit**, also warrants clarification. In particular, I want to highlight its use in identifying explanatorily irrelevant parameters. Taking a function to an asymptotic limit simply means identifying how the function behaves when one or more of its arguments approaches either zero or infinity. Let's start with a toy example. Consider the function $f(x) = x^2 + 9$. Clearly:

$$\lim_{x \to 0} f(x) = 9.$$

As this shows, x becomes moot in the limit. The general goal of asymptotic explanations is to get (a mathematical representation of) the **pattern of behavior** in the aforementioned schema to be what emerges in the limit. If it's useful, think of asymptotic limits' explanatory benefits as a hypothetical set of "extreme conditions." These extreme conditions help abstract away the many explanatorily irrelevant factors that are present in normal conditions. If a certain explanatory factor can be found even in these extreme conditions, it's liable to be present (to some degree or another) in all other systems of interest. Consequently, its relevance to the explanandum is warranted.

Many of Batterman's (2002, 16) examples of asymptotic reasoning utilize *dimensional analysis*. In this context, dimensional analysis involves examination and manipulation of the various dimensions involved in a given problem, with the goal of creating dimensionless parameters. By taking one or more of these dimensionless parameters to be either very small or very large and implementing the appropriate limit, a constant can replace one of the parameters, thereby yielding equations more tractable than the original. Often, these equations exhibit self-similarity, or scaling behavior. Importantly, self-similarity indicates that a type of stability has been achieved independently of details regarding initial and boundary conditions. Even in the frequent cases where such limits don't exist, more complex mathematics can be implemented and self-similarity can emerge.[8]

Bjorken offered an asymptotic explanation of (E2) using dimensional analysis. In his account of inelastic scattering, Bjorken initially relied on effective mass, which is a dimensional constant. As noted earlier, dimensionless quantities are quite useful in cases where asymptotics are to be used. Accordingly, Bjorken created the dimensionless *Bjorken scaling variable*, $\omega = -q^2 / Mv$. This tool proved handy, as it led him to the "Bjorken limit," in which v (the energy lost by the electron in collision) and q^2 (the

---

[8] See Batterman (2000, 254; 2002, 16) for further details.

square of the momentum transfer between the electron and the proton) are both taken to infinity, but the ratio $\omega = v/q^2$ is held constant.

Thus, the limits in question were:

$$\lim_{q^2 \to \infty, \frac{v}{q^2} fixed} vW_2(q^2, v) = F_2(-q^2/Mv) = F_2(\omega)$$

and

$$\lim_{q^2 \to \infty, \frac{v}{q^2} fixed} MW_1(q^2, v) = F_2(-q^2/Mv) = F_1(\omega).$$

For the mathematically disinclined, you can steal a glimpse at Bjorken's central theoretical insight by observing the punctuation in the argument of the Bjorken functions F. The division sign signals that, initial appearances notwithstanding, q^2 and v *depend on each other* in determining the values of the structure functions – a point that can be observed only when they're approaching "extreme conditions" (i.e., when both approach ∞). Take an especially close look at "$q^2 \to \infty$." This means that we're at very high energies. The rest of the equation says that, despite these high energies, the proton's structure (represented by W_1 and W_2) becomes a function of the ratio (ω) between the square of the momentum transfer between the electron and the proton (q^2) and the electrons' energy loss (v).

With these ideas in hand, we can see how Bjorken instantiated the asymptotic explanatory pattern in his explanation of scaling; i.e., (E2):

Explanation Target
Why does $W_1 = F_1(\omega)$ and $vW_2 = F_2(\omega)$ in diverse lepton-hadron scattering experiments?
Application of Explanatory Pattern
W_1 and W_2 can be expressed as mathematical functions of q^2 and v.
E and θ are constant in the Bjorken limit in W_1 and W_2.
Given the underlying microphysics of the lepton–hadron scattering experiments, differences in E and θ are irrelevant to W_1's equaling $F_1(\omega)$ and vW_2's equaling $F_2(\omega)$ in the lepton–hadron scattering experiments[9].

---

[9] Importantly, it's debatable if Bjorken provided the **underlying microphysics**. I address this in Section 2.3.1.

Here, E refers to the beam energy of the electrons and θ refers to the scattering angle of an outgoing electron. Importantly, others previously thought that E and θ are or could be explanatorily relevant to the structure functions.

Because W_1 and W_2 determine σ_{DIS}, there is an intimate link between the scaling phenomena (E2) and the other explanandum (E1), the unexpected scattering results. More precisely, it was well known prior to the scattering experiments that:

$$\sigma_{DIS} \approx \sigma_{MOTT} \left[W_2 + 2W_1 tan^2 \left(\frac{\theta}{2}\right) \right].$$

Here, θ refers to the scattering angle of an outgoing electron. Effectively, Bjorken's scaling laws imply that the expression

$$W_2 + 2W_1 tan^2 \left(\frac{\theta}{2}\right)$$

approaches 1 as we increase the energy level of the scattering experiment. But this means that $\sigma_{DIS} \approx \sigma_{MOTT}$ $\sigma_{DIS} \approx \sigma_{MOTT}$, which is equivalent to (E1). Thus, Bjorken could easily explain the deep inelastic scattering results as a consequence of his explanation of scaling.

2.3.   A Lack of Understanding

Clearly, Bjorken advanced our understanding of these scattering results and scaling laws. The Nexus Principle tells us why: he provided us with new (and correct) explanatory information. Nevertheless, when the Bjorken scaling curves were first discovered, experimenters did not understand why (E1) and (E2) were true. Indeed, as the SLAC-MIT group's paper reporting of the scaling phenomenon was being written for publication, one of Bjorken's postdocs, Emmanuel Paschos, was quoted as saying, "The experimenters have this puzzling graph of the structure function BJ asked them to make . . . He claims the data should 'scale' and it does, but nobody seems to *understand* what this 'scaling' means" (Riordan 1987, 150; emphasis added).[10]

Note that here we must appeal to our analysis of outright understanding from Chapter 1:

---

[10] Cao (2010, 96, 98) and Pickering (1984, 132) also argue that Bjorken's account was not well understood by other physicists.

Outright Understanding: "S understands why p" is true in context C if and only if S has minimal understanding and S approximates ideal understanding of why p closely enough in C.

While Bjorken provided an explanation of the phenomena, he did not hit the contextually relevant benchmarks. What more did the other physicists want? Two desiderata suggest themselves. First, it's debatable if Bjorken filled in one variable in this explanation schema – namely, the **underlying microphysics** of the scattering experiments. Thus, more explanatory information was needed, just as the Nexus Principle suggests (Section 2.3.1). The positive reception of Feynman's subsequent developments of Bjorken's explanation suggests that this was an important factor. Second, according to the Scientific Knowledge Principle, *having* an explanation is one thing, but understanding improves when one has *scientific knowledge* of that explanation (Section 2.3.2). The response to Bjorken's asymptotic explanation suggests that physicists were not convinced that there was sufficient experimental evidence for Bjorken's explanation.

Before turning to this history, I want to emphasize two things. First, I stress that Bjorken's shortcomings shouldn't be overstated. He provided many of the important details of the correct explanation of the scattering results and the scaling curves. Thus, per the Nexus Principle, our understanding of these phenomena is *better* as a result of his work. However, this doesn't mean that we should credit him with providing his peers with the *outright* understanding that they were searching for. Indeed, we'll now see how subsequent physicists gained an *even better* understanding than what Bjorken had provided in the years following his initial discoveries. Second, at this stage of the narrative, my goal is to present a plausible story in which the EKS Model accounts for salient features of the physicists' practices. I will fend off objections and potential counter-narratives in Sections 2.4 and 2.5.

2.3.1. Incomplete Explanation

Let's examine the first factor that prevented experimental physicists from understanding the scaling phenomena: the incompleteness of Bjorken's explanation. Paschos's exasperation about the opacity of Bjorken's explanation fell upon the ears of Richard Feynman, who expanded on that explanation with the "parton" model in August 1968. The core idea is that (*pace* QED) hadrons are composed of hard, point-like entities that Feynman called partons (and which are now called quarks.)

Although it was formally quite similar to Bjorken's current-algebraic account of scaling, the parton model added a mechanical interpretation that experimenters found easier to understand than Bjorken's dense mathematics. Indeed, when Bjorken first made the scaling prediction, he had entertained the possibility that hadrons were composed of quarks but was not particularly enamored with the idea, later describing it as "the most trivial, simple representation of local current algebra that you could think of."[11] However, thinking of hadrons in this way turned out to be a pivotal turning point in physicists' understanding of these phenomena. One MIT experimenter remarked that with the advent of Feynman's parton model, "Experimenters could finally talk to theorists in a language both understood" (Riordan 1987, 152).

Before proceeding, let's elaborate the sense in which Bjorken's explanation was incomplete. First, "incompleteness" should not be understood here to entail some failure to meet requirements of an ideal explanation. Rather, it is a failure to provide values for all of the variables involved in an accepted or acceptable explanation schema, as is the case with Bjorken's failure to fill in the Asymptotic Explanation Schema described earlier.

Second, the explanation is incomplete because **underlying microphysics** is a variable in the Asymptotic Explanation Schema, but it would be hasty to infer from this single historical example that all understanding-conferring explanations must be microphysical, reductive, or mechanistic in character. Indeed, the requirement for an underlying microphysics comes from Batterman (see note 7), who explicitly denies that asymptotic explanations are mechanistic (Batterman 2002, 9–13).

As a result, Feynman's contribution was valuable precisely because it provided a mechanical model that filled out important gaps in Bjorken's explanation. Feynman achieved this by deploying two strategies previously used while theorizing about high-energy proton–proton interactions. These ideas then readily applied to both scaling and deep inelastic electron–proton scattering.

The first strategy is called "working in the infinite momentum frame." Feynman imagined that, because of their high relative velocity, each proton would "see" the other as relativistically contracted along its direction of motion – roughly as a "pancake." Because the strong interactions between protons are of short range, each proton would also see the other as a frozen

---

[11] As reported in Riordan (1987, 153). Similarly, Cao (2010, 89) reports that "Bjorken was not particularly fascinated by the constituent quark model for various reasons [which] made it easier for him to move away from the constituent quark model as an underpinning picture in his further explorations."

snapshot of its constituent particles. At this stage of research, Feynman was agnostic as to whether these particles were quarks or other hadrons, so he simply used the term "partons" as a placeholder. Using the second strategy – impulse approximation – he further assumed that within a single pancake (i.e., proton), partons did not interact with each other. Consequently, in strong interactions, each parton acts as an independent, quasi-free entity.

This picture of the proton enabled Feynman to complete Bjorken's two aforementioned explanations. First, in deep inelastic scattering, the incoming electron emits a photon, which then interacts with a single free parton. Much like electrons, Feynman assumes that partons are structureless and point-like. Consequently, the cross-section of protons in the deep inelastic region is similar to the cross-section of electrons in this region (i.e., $\sigma_{DIS}/\sigma_{MOTT} \approx 1$ (E1)).

Second, Feynman explained scaling as follows: W_1 and W_2 measure the distribution of the partons' momentum within the proton. In an interaction between an electron and a proton, the partons' momentum would be determined entirely by the momentum transfer between the electron and the proton (q) and the electron's energy loss (v). Consequently, W_1 and W_2 scale with functions of ω (E2).

Thus, Feynman's mechanical parton model filled in much of the **underlying microphysics** that Bjorken's explanation lacked. If one likes, this functions as a "higher-resolution module" in the Asymptotic Explanation Schema:

> *Mechanical Model Schema*
> *Explanation Target*
> How do the **underlying microphysics** affect a **pattern of behavior** in a **system**?
> *Explanation Pattern*
> The **system** is made up of **parts**.
> The **parts** have **properties** and **interact** with each other.
> According to the **underlying microphysics, interacting** objects with these **properties** are subject to certain **laws**.
> The **pattern of behavior** is a **product** of these **law**-governed **interactions**.

Feynman filled in the schema as follows:

> Explanation Target
> Why does $W_1 = F_1(\omega)$ *and* $vW_2 = F_2(\omega)$ in an electron-proton interaction?

Application of Explanation Pattern

The electron–proton interaction is made up of electrons and partons.

The electrons and partons are hard and point-like and collide with each other.

Per the infinite momentum frame and impulse approximation, when hard and point-like objects collide, their momentum transfer is a function of ω.

W_I and W_2 are momentum distributions of the partons in these collisions.

Moreover, Bjorken and Feynman's collective explanation is correct – which is a requirement of the EKS Model. By the standards of current science, partons are the theoretical precursors to quarks. Historically, it's worth noting that the quark idea was alive and well by 1964 – several years before the SLAC-MIT experiments. However, the lack of any empirical evidence of any free-standing, fractionally charged quarks made them unpopular with many particle physicists throughout the early 1970s. Undoubtedly, this partly motivated Bjorken and Feynman to adopt a more ontologically modest stance that, for example, leptons' colliding with *something point-like* in a hadron correctly explains the deep inelastic scattering and scaling phenomena. Hence, we can see Feynman's contribution as an application of the Nexus Principle: he provided further explanatory information about the scattering results and the scaling laws.

Zooming out a bit, we can see how Bjorken and Feynman's explanation fits the broad account of explanation presented in Chapter 1. There I claimed that q explains why p just in case:

(1) p is (approximately) true;

(2) q makes a difference to p;

(3) q satisfies your ontological requirements (so long as they are reasonable); and

(4) q satisfies the appropriate local constraints.

Given that the explananda, (E1) and (E2), are known, they are true. So, the first condition is satisfied. The third condition is also readily satisfied: Feynman's partons eventually became the quarks of the standard model of particle physics. Depending on one's stance toward scientific realism, this model is either empirically adequate or approximately true.

However, the second and fourth conditions stand in an interesting tension. The local constraints in this explanation are encapsulated in the Asymptotic Explanation Schema and the Mechanical Model Schema. The first of these emphasizes that many things *don't* make a difference to the behavior described in the explananda and thus seems at odds with the second condition, which requires an explanans to make just such a difference. There are, however, two ways of resolving this tension.

First, and more contentiously, we might see a kind of "second-order difference-making" in play with asymptotic reasoning. Consider a more mundane, "first-order" example of difference-making:

> Had John taken a lower dosage, he would not have recovered.

Note that asymptotic reasoning does *not* support the following first-order difference-making claim:

> Had the details *been different*, then the same pattern of behavior would not have emerged in diverse physical systems.

The entire point of asymptotic analysis is to belie such claims: the details don't matter.

Contrast this with the kind of "second-order" difference-making that asymptotic explanations do support:

> Had the details *made a difference*, then the same pattern of behavior would not have emerged in diverse physical systems.

I call this second-order difference-making because the antecedent of the counterfactual makes explicit reference to difference-making, which isn't necessary in more mundane, first-order cases of difference making. The rough idea is this: that the details don't make a difference makes a difference to the universal behavior being explained.

As I said, I consider this contentious (and a little mind-bending), so I want to offer a second reason for reconciling asymptotic explanations with difference-making that skates freely of second-order difference-makers. The fact that many details about the different systems don't make a difference to the behavior in question is consistent with other details (what I called the "underlying microphysics") making a difference.[12] Indeed, this is precisely what we find in our example. Bjorken's explanation mostly highlights the details that don't make a difference (beam energy and scattering angle), while Feynman's highlights the microphysics that does (namely partons and their interactions with electrons). This is perhaps clearest when we examine the alternative explanations of (E1) and (E2),

---

[12] For more on the importance of integrating non–difference-making and difference-making details in asymptotic explanation, see Batterman and Rice (2014).

where alternative models of proton structure yielded false predictions, thereby showing where the fact that partons constitute protons makes a difference; so let's now look at these alternative accounts of the fine structure of the subatomic world.

2.3.2. *Explanations UnSEEn*

Thus far, we've seen that Bjorken's explanation did not provide understanding to other physicists because it was incomplete, but that Feynman's mechanical parton model filled this gap. The EKS Model also suggests that merely *possessing* a correct explanation frequently falls short of understanding – the explanation must be *known*, to wit in a way characteristic of scientific practice. Earlier, I characterized scientific explanatory evaluation or SEEing as a common form of scientific knowledge about explanations. As I'll now show, after Feynman's interpretation, experimental physicists engaged in SEEing primarily by running additional experiments designed to rule out competing (i.e., other plausible potential) explanations of the deep inelastic scattering and scaling phenomena.

Recall SEEing's three features: the consideration of plausible potential explanations, comparative assessment of those explanations, and formation of doxastic attitudes on the basis of those assessments. Even in the *early* stages of this historical example, the particle physicists in question evaluated explanations in roughly this way. QED offered one potential explanation of the scattering experiments. If it were correct, then σ_{DIS} would have been low. But the experimental evidence indicated the opposite. Similarly, if the high σ_{DIS} were a result of photon radiation, then σ_{DIS} would have been low after the radiative corrections. However, this also turned out to be false. So Bjorken's current-algebraic model gained traction.

Yet, the extent to which Bjorken's explanation was produced via SEEing is limited and, importantly, was also unclear to his contemporaries. This is supported by the fact that further explanations were considered and compared to the parton model.

Specifically, two other models offered potential explanations of these phenomena. Each of these explanations was rejected using the same process of explanatory evaluation just described. Importantly, the experimental physicists designed and conceived of experiments that ruled out these explanations only after learning of Feynman's addenda. Hence, per the EKS Model, Feynman's amendments to Bjorken's explanation contributed to understanding not only by adding to our stockpile of correct explanatory information, but also by facilitating scientific explanatory evaluation.[13]

First, in August 1969, the SLAC group, this time led by Richard Taylor, designed another series of experiments intended to adjudicate between a parton model and Sakurai's vector meson dominance model. Sakurai held that proton–electron interactions were mediated by "vector mesons" instead of Feynman's partons. The vector meson model explained the scaling phenomenon. According to Sakurai, $R = \sigma_L/\sigma_T$, the ratio of the proton's tendencies to absorb virtual photons longitudinally to its tendencies to absorb those photons transversely, should be quite large (in between 1 and 10). By contrast, parton models held that it should be quite small (in between 0 and 1). The experiments strongly indicated the latter, and Taylor's presentation of their results at the Electron-Photon Symposium held in Liverpool during September 1969 is widely regarded as the death-knell of vector meson dominance theories, which provided the most widely accepted explanation of scattering behavior in the high-energy physics community for most of the Sixties.

Similarly, in 1971–1972, experimenters at the European Center for Particle Physics (CERN) conducted experiments highlighting advantages of the parton model over Arbanel et al.'s Regge exchange explanation of scaling. According to the Regge exchange model, scaling was the result of a whole series of hadrons being exchanged during the electron–proton collision. The Regge exchange model posited a soft hadron. Thus, it predicted that collisions between hadrons would produce few particles at larger angles. However, when CERN ran proton–proton collisions, it found a far larger number of particles at these large angles than the Regge exchange model could account for. In contrast, the parton model explained this phenomenon with relative ease: protons are composed of small structures that go ricocheting off one another during such collisions.

In evaluating these explanations and designing these experiments, the physicists introduced further explananda, e.g.:

(E3) *R* is below (rather than above) 1.

(E4) Many (rather than fewer) particles were found at large angles in the proton–proton scattering experiments at CERN.

Moreover, as with (E1) and (E2), the experimenters took measures to guarantee that these phenomena were not mere artifacts. Thus, the physicists had *knowledge that* these explananda are true.

---

[13] Admittedly, this is the most contentious part of my narrative. I address some objections to it in Sections 2.5.2 and 2.5.5.

Additionally, the physicists were clearly evaluating explanations using reliable experimental methods. Hence, their belief that the parton model correctly explained the deep inelastic scattering and scaling phenomena could not easily have been false. Consequently, the requisite cognitive processes produce their true beliefs safely. So, they possessed the kind of explanatory knowledge that the EKS Model equates with understanding.

2.4. Comparison with Batterman

So where are we? I've shown that by the time the parton model was accepted as providing understanding of (E1) and (E2), physicists had true beliefs about the correct explanations of these two phenomena, and these beliefs were supported by scientific explanatory evaluation. In other words, the EKS Model provides a plausible interpretation of the physicists' norms of understanding. However, perhaps there are other plausible interpretations of this scientific episode. To conclude this discussion, let's look at two contenders. In this section, I compare the EKS Model to Batterman's account of understanding. In the next, I compare it to De Regt's account of understanding.

As we've already seen, Batterman's account of asymptotic explanation provides a productive framework for interpreting Bjorken's reasoning. Using this framework, Batterman has explicitly avowed an account of scientific understanding:

> The explanations I have been discussing involve, essentially, methods for extracting structurally stable features of the equations that purportedly model or govern the phenomenon of interest. These features are often emergent in appropriate asymptotic domains, and their perspicuous mathematical representation constitutes an essential component of our understanding of the world. (Batterman 2002, 59)

As Batterman does not take his view to provide a comprehensive account of scientific understanding, it is best seen as a limiting case of the EKS Model. First, Batterman's account of understanding puts special emphasis on asymptotic explanations. As Batterman would agree, this surely is not a requirement for all understanding, for it would imply that scientific understanding is necessarily mathematical and, moreover, must always involve asymptotic reasoning. By contrast, the EKS Model allows for non-mathematical understanding in the sciences. Consider, for instance, Semmelweis's oft-discussed discovery of the causes of childbed fever, for example (Lipton 2004). Semmelweis accepted a (broadly) correct yet qualitative explanation of childbed fever via the aforementioned process of explanatory evaluation.

Indeed, even Feynman's contribution above suggests that non-mathematical explanations frequently complement asymptotic ones. Thus, as the EKS Model suggests, not only asymptotic explanations provide understanding.

The EKS Model also clarifies a lacuna in Batterman's account of understanding. Batterman does not specify the cognitive relationship whereby an asymptotic explanation provides understanding to a scientist. Simply "having" an asymptotic explanation does not guarantee that it provides understanding. For instance, the experimental physicists clearly "had" Bjorken's asymptotic explanation prior to Feynman's amendments in some sense, yet the preceding quotations indicate that this explanation afforded them little understanding. Rather, they understood the phenomena only when they could SEE the explanation of those phenomena through experiments that ruled out Sakurai and Regge's models. This is precisely what the EKS Model suggests.

2.5. Comparison with De Regt

De Regt provides another prominent view of understanding. After presenting his position, I'll argue that the EKS Model has several advantages over it. De Regt's two core commitments are a Criterion for Understanding Phenomena:

> (CUP) A phenomenon P is understood scientifically if a theory T of P exists that is intelligible (and the explanation of P by T meets accepted logical and empirical requirements),

and a Criterion for the Intelligibility of Theories:

> (CIT) A scientific theory T (in one or more of its representations) is intelligible for scientists (in context C) if they can recognize qualitatively characteristic consequences of T without performing exact calculations. (De Regt 2009b: 32–33)

Combined, these entail De Regt's Account of Understanding:

> (DRAU) A phenomenon P is understood scientifically if a theory T of P exists such that:
> (1) Scientists (in some context C) can recognize qualitatively characteristic consequences of T without performing exact calculations; and
> (2) The explanation of P by T meets accepted logical and empirical requirements.

To its credit, DRAU nicely captures how Feynman's parton model provides understanding. Furthermore, De Regt's view is sufficiently rich that I can't subsume it under the EKS Model, as I did with Batterman's view. Nevertheless, let's consider several advantages that the EKS Model enjoys over De Regt's account.[14]

2.5.1.   Better Coverage

My view covers more instances of understanding than De Regt's. For instance, sometimes scientists come to understand things through the use of quantitative and formal vocabularies. As an example, consider the Asymptotic Explanation Schema, where the **asymptotic limit** variable is typically determined through exact calculation. This accounts for much of Bjorken's contribution to our understanding of the phenomena described earlier. Furthermore, we've seen that the EKS Model accounts for this. By contrast, DRAU accords no role to quantitative reasoning in scientific understanding.

Admittedly, since De Regt does not specify necessary conditions for understanding, he need not deny that scientists can gain understanding in this way. Perhaps there is some other way to understand a phenomenon with exact calculations that is not captured by DRAU. However, he has not offered an explicit account of this kind of understanding. By contrast, the EKS Model provides a unified account of both qualitative and quantitative understanding.

2.5.2.   Understanding as a Scientific Goal

Intuitively, understanding is both worth having and an aim of science. The EKS Model suggests that the point of understanding is to believe correct explanations (however "correct explanations" are glossed in the final analysis), which sits well with the shopworn claim that explanation is an important scientific activity.[15] Scientific explanatory evaluation is thereby instrumentally valuable (i.e., an effective means to believing only correct explanations).

Like me, De Regt (2009b, 26) grants that "Having an appropriate explanation ... is ... an essential epistemic aim of science." However, unlike me, he then argues that this involves the kind of understanding described by DRAU. Yet my previous point against DRAU (Section 2.5.1) implies that understanding is not limited to qualitative reasoning in the absence of exact calculation. Consequently, if understanding is valuable, it's not always because of the features showcased in DRAU.

---

[14] I provide a more extensive critique of De Regt in Khalifa (2012).

[15] Note that even if something like Van Fraassen's account is right, then explanations can still be epistemically valuable, albeit indirectly: "the *epistemic* merits a theory may have or must have to figure in good explanations are not *sui generis*; they are just the merits it had in being empirically adequate, of significant empirical strength, and so forth" (van Fraassen 1980, 88).

To that end, the EKS Model identifies the conditions wherein qualitative reasoning in the absence of exact calculation contributes to the value of understanding. Let us return to an earlier quote from Michael Riordan, one of the MIT researchers at the time:

> Experimenters could finally talk to theorists in a language both understood . . . Feynman had again supplied a language, a strikingly simple mental image, to describe what might be going on in a remote and tiny realm. (Riordan 1987, 152)

Such a quote sits nicely with DRAU. But *why* is it so important that experimenters and theorists have a *lingua franca* or "strikingly simple mental images"? Suppose that experimenters and theorists continued to talk past each other. For instance, no matter how much the experimenters studied their current algebra, it just didn't have any traction in the way they designed their experiments. Quite plausibly, theorists would offer potential explanations, but experimenters would be unreliable in providing relevant evidence to test those hypotheses. In short, the experiments that squarely ruled out the vector meson dominance model and the Regge exchange model would be either poorly designed or never designed in the first place. Scientists would be facile with linguistic or imagistic tropes but would regularly fail to believe in correct explanations. Clearly, their understanding of relevant phenomena would suffer.

This suggests that expressive languages, fecund visual models, and qualitative reasoning more generally contribute to understanding only insofar as they allow scientists to evaluate explanations more effectively. In other words, we value the kind of qualitative reasoning that stands at the heart of DRAU only insofar as it allows us to SEE the nexus (i.e., to have scientific knowledge of an explanation). Stated a bit more baldly, the value of qualitative insight (a shared language, visual images, etc.) is exhausted by its facilitation of good old-fashioned hypothesis testing.

2.5.3. *The Problem of Irrelevant Insights*

The preceding considerations were of a general flavor. I'll now tout some of the Nexus Principle's virtues relative to De Regt's account. First, as formulated, one can "disconnect" the two conditions of DRAU's antecedent to produce fairly straightforward counterexamples to it. In particular, scientists may be able to recognize another qualitatively characteristic consequence Q of T that has nothing to do with P, but fail to recognize that T explains P. According to DRAU, this suffices to generate understanding of P. That seems quite doubtful. Let's call this the *problem of irrelevant insights*.

Consider an example: in the aftermath of the April 2010 British Petroleum oil spill in the Gulf of Mexico, a Berkeley research team reported the discovery of a new species of "oil-eating" deep-sea psychrophilic gammaproteobacteria in late August 2010 (Hazen et al. 2010). Undoubtedly, there was a point in recent history (e.g., before the oil spill) where there were many unrecognized consequences of our current microbiological theory concerning this particular species of gammaproteobacterium (e.g., that it does not retain crystal violet in the Gramstaining protocol).

However, at the same time that *these* consequences remained unrecognized, scientists could easily recognize that *Streptococcus pneumoniae is a bacterium* is a consequence of the same theory. According to DRAU, simply because scientists can recognize that *Streptococcus pneumoniae is a bacterium*, it follows that *deep-sea psychrophilic gammaproteobacteria do not retain crystal violet in the Gram-staining protocol* can be understood. This seems wildly implausible.

By contrast, the Nexus Principle dissolves the problem of irrelevant insights because no model of explanation would claim that *Streptococcus pneumoniae is a bacterium* explains (and hence provides understanding of) why *deep-sea psychrophilic gammaproteobacteria do not retain crystal violet in the Gram-staining protocol*.

2.5.4.   The Problem of Improbable Explananda

Since the consequences of the problem of irrelevant insights seem very implausible, I assume that De Regt intends something other than DRAU. The most plausible proposal would replace (1) in DRAU with:

> (1')   Some scientist S (in some context C) can recognize P as a qualitatively characteristic consequence of T without performing exact calculations.

This would block the problem of irrelevant insights. Since the scientists did not recognize anything about gammaproteobacteria, they did not understand anything about these bacteria, either.

Furthermore, this interpretation is consistent with De Regt's discussion of Boltzmann's understanding of the macroscopic properties of gases via the kinetic theory:

If one adds heat to a gas in a container of constant volume, the average kinetic energy of the moving molecules – and thereby the temperature – will increase. The velocities of molecules therefore increase and they will hit the walls of the container more often and with greater force. The pressure of the gas will increase. In a similar manner, we can infer that, if temperature remains constant, a decrease of volume results in an increase of pressure. Together these conclusions lead to a qualitative expression of Boyle's ideal gas law. (De Regt and Dieks 2005: 152)

The phenomena to be understood are the relationships among temperature, pressure, and volume expressed by the ideal gas laws. The understanding consists of inferring a qualitative version of these laws from a qualitative formulation of the kinetic theory, as would be the case under (1').

This proposed solution to the problem of irrelevant insights is tentative, for it provides the seeds of De Regt's undoing. Note that if he shifted from (1) to (1') in DRAU, he would be on the hook for the following claim about explanation:

- If T explains P, then P is a consequence of T.

However, the resulting view about explanation faces some venerable problems for it is one of many Hempelian ideas to face searching counterexamples. The most famous example on this front is that a person's having syphilis explains why he has paresis, yet only 25 percent of syphilitics suffer from paresis (Scriven 1959). As a result, one could not recognize paresis as a consequence of a theoretical claim about syphilis. Analogous examples appear throughout the special sciences. For example, in neuroscience, as little as 10 percent of all action potentials result in neurotransmitter release, but action potentials are regarded as the central causal mechanisms explaining neurotransmitter release (Bogen 2005; Craver 2007).

Surprisingly, De Regt does not address these sorts of examples despite the challenge they pose for his view. Given that he openly acknowledges the strong similarities his view bears to Hempel's deductive-nomological model and addresses several well-trodden problems concerning the *sufficiency* of that account, such as the barometer and flagpole problems (De Regt and Dieks 2005: 162–163), ignoring the signal challenge to the *necessity* of Hempel's model is a significant oversight.

Moreover, De Regt's handling of the other Hempelian counterexamples does not easily transfer to this case. For instance, when handling explanatory asymmetries, he follows van Fraassen (1980) in claiming that "it depends on the context whether the length of the flagpole makes it understandable how long the shadow is, or vice versa" (De Regt and Dieks 2005: 164). Whatever the merits of that approach, appealing to context in the syphilis example involves swallowing a bigger pill because the challenge only requires that if T explains P, then P is a consequence of T. This doesn't require reference to context: the consequence relation is an issue of semantics, not pragmatics. If De Regt seeks to challenge that claim, then he is shouldering a rather significant burden of proof in the philosophy of logic.

Furthermore, even if De Regt bit the bullet and insisted that the consequence-relation *is* context-sensitive, he's surely inviting the charge that understanding is epistemically suspect. If, in a particular context, one can "infer" that someone has paresis from the fact that he has syphilis despite the low conditional probability, then recognizing a theory's consequences is little more than forming psychological associations with that theory. De Regt would then seem hard pressed to reconcile this with his claim that understanding is not "merely a (philosophically irrelevant) psychological by-product of scientific activity" (De Regt and Dieks 2005: 138).

As before, the Nexus Principle provides a tidy dissolution of this problem. A person's having syphilis *causes* him to have paresis, and this is widely regarded as the relevant notion of explanation in this example even if there is no further inferential relation.

2.5.5. The Epistemology of Understanding

So far, I've advertised both general attractions of the EKS Model and also extoled the virtues of the Nexus Principle in comparison to DRAU's condition (1). I think that the Scientific Knowledge Principle is also an improvement over condition (2) of DRAU, which states that understand-ing-conferring explanations must satisfy "the accepted logical and empiri-cal requirements."

It's worth noting how little conflict there needs to be here. In particular, since DRAU only states that these logico-empirical requirements are *sufficient* for understanding, I'll have nothing to say about cases of under-standing in which explanations do not satisfy these requirements since De Regt can always claim agnosticism about how these cases should turn out. Additionally, there will be many instances in which SEEing and De Regt's accepted requirements are coextensive. Indeed, as long as the "accepted logical and empirical requirements" track with the best scientific methods available, De Regt and I will be in substantial agreement.

Nevertheless, even if this rosy picture turns out to be correct, De Regt and I will still have a minor quarrel, for I'll insist that safe beliefs do some important work in our understanding; De Regt appears to have no such commitment. I'll first argue that he *may* be more committed than his position (i.e., DRAU) suggests, and, failing that, he *should* be more committed than his position suggests.

First, I want to question whether it's even possible that scientists could have easily formed the wrong doxastic state about an explanation while nevertheless satisfying De Regt's "accepted logical and empirical considerations." On such a scenario, although T correctly explains P in the actual world, T could have easily incorrectly explained P on the basis of the same logical and empirical requirements. This proposal is potentially incoherent: in science, the accepted logical and empirical considerations typically aren't ones that permit such scenarios because if an explanation could have easily been false given the logical and empirical criteria it satisfies, then one should not accept those criteria but should instead seek out more demanding criteria that will rule out false positives. For instance, if the logical and empirical criteria were this lax, De Regt's account would leave it rather mysterious as to why our scientists performed experiments that ruled out the vector meson dominance and Regge exchange models. Consequently, it may be that, upon closer inspection, the accepted logical and empirical requirements must be requirements that deliver safe beliefs in explanations. Lest this all seem too abstruse, this boils down to the commonsense idea that our understanding is better when it's grounded in solid evidence than when it's grounded in shaky evidence. If all of this is correct, then De Regt and I are both steadfast proponents of the Scientific Knowledge Principle.

However, let's suppose that De Regt claimed that there is some sense in which an explanation can satisfy his empirical and logical requirements, yet the relevant explanatory propositions could easily have been false. While such a case satisfies DRAU without satisfying the EKS Model, the understanding it licenses is dubious. Specifically, understanding could be achieved by having an explanation that could have easily been incorrect given the available evidence; e.g., so long as scientists could make the requisite qualitative inferences with the parton model, their inability to eliminate its rivals (vector meson dominance, Regge exchange) matters little. However, this implies that understanding of (E1) and (E2) is achieved even if one can only draw qualitative consequences from one of these rivals since they also satisfy these (weakened) empirical and logical requirements.[16] So abandoning the Scientific Knowledge Principle in this way undermines DRAU's plausibility.

---

[16] In subsequent work, De Regt (2015) seems to embrace this consequence. I discuss that work in Chapter 6.

Now, De Regt might reply to this by insisting that understanding is one thing and confirmation is another. For instance, one might claim that, in the events described earlier, the real moment of understanding ends with Feynman's innovations and that the experiments that follow are distinct from understanding. However, this puts further pressure on De Regt to address our earlier point about the value of understanding – why would qualitative reasoning with a theory that satisfies *fairly weak* logical and empirical constraints be something worth having?

Thus, all told, De Regt's view seems to be less comprehensive than the EKS Model, and it fails to account for understanding's status as a scientific goal as well as the EKS Model. By departing from the Nexus Principle, it suffers from the problems of irrelevant insights and improbable explananda. If it departs from the Scientific Knowledge Principle, then its logical and empirical constraints are too modest to discriminate between correct and incorrect explanations.

2.6. Conclusion

I have presented a simple model of understanding – the EKS Model. I then showed how this model accounts for an episode in the history of science which functions as a kind of natural experiment for theories of scientific understanding. Moreover, I've argued that the EKS Model has certain advantages over two leading alternatives, Batterman's and De Regt's.

I hope that my epistemologist friends aren't left cold by this exercise. On the one hand, this shows that epistemological ideas aren't merely the stuff of armchair reflections, as we've seen that epistemological concepts such as safety play an important role in the philosophical retelling of Bjorken's story. Additionally, for philosophers of science, the deployment of epistemological concepts can help to adjudicate between relatively kindred views, such as my own and De Regt's. So maybe there's hope for further crosstalk between epistemology and philosophy of science going forward.

Let's also not forget the larger dialectic at stake here. For proponents of the received view – as well as my EKS Model – figuring out how explanations are known renders any further philosophical theorizing about understanding largely gratuitous. We've just worked our way through a fairly sophisticated instance of understanding using only the Nexus and Scientific Knowledge Principles. While I obviously can't rest my whole case on a single example, I hope to have shown how an epistemology of explanation need not be modest in its resources, even if it is modest in its philosophical ambitions.

CHAPTER 3

Understanding and Ability

Understanding frequently involves impressive exercises of cognitive ability. For instance, looking back to Chapter 2, Bjorken and Feynman's theoretical insights were nothing short of remarkable. Similarly, the experimenters who vindicated those insights deployed impressive skills and astonishing technological know-how. Even in more mundane, non-scientific contexts, many take understanding's most distinctive feature to be a certain kind of ability to "grasp" or "see" how things hang together.

Clearly, knowledge requires nothing this fancy; yet, as you'll recall from the first chapter, the received view states that understanding just *is* knowledge of an explanation; i.e.:

> S understands why *p* if and only if there exists some *q* such that S knows that *q explains why p.*

Hence, the possibility that understanding has some abilities that evade knowledge prompts the Classic Ability Question:

- Do understanding and knowledge require the same abilities?

Proponents of the received view answer this affirmatively; their critics, negatively.

Furthermore, recall that the received view is my inspiration but not my final word on understanding. In its place, I proposed the Explanation-Knowledge-Science (EKS) Model, which states:

> (EKS1) S_1 understands why *p* better than S_2 if and only if:
> (A) *Ceteris paribus,* S_1 grasps *p*'s explanatory nexus more completely than S_2; or
> (B) *Ceteris paribus,* S_1's grasp of *p*'s explanatory nexus bears greater resemblance to scientific knowledge than S_2's.[1]

---

[1] Recall the implicit proviso that S_1 also has minimal understanding of why *p*.

(EKS2) *S* has minimal understanding of why *p* if and only if, for some *q*, *S* believes that *q explains why p*, and *q explains why p* is approximately true.

Because the EKS Model gives degrees of understanding their proper due, the Classic Ability Question ends up being a bit off-point. On my view, some modest instances of understanding require *less* in the way of ability than the received view. For this reason, I first clarify and motivate the Classic Ability Question (Section 3.1) and then show that a negative answer is unfounded (Section 3.2).

On the other hand, certain grades of understanding require more in the way of ability – namely the abilities characteristic of *scientific* knowledge – than the received view. This suggests a different question about understanding's relationship to abilities. In Chapter 1, I called this the *Updated Ability Question*:

• Which abilities, if any, improve our understanding?

So, after making my peace with the Classic Ability Question, I offer my own answer to the Updated Ability Question (Section 3.3). Using three alternative answers to this question as my foils (Sections 3.4–3.6), I argue that the abilities that improve our understanding are precisely those that furnish us with scientific knowledge of an explanation. Consequently, the EKS Model outperforms this trio of contenders.

3.1. The Classic Ability Argument

When I first proposed the idea that understanding is explanatory knowledge, the Classic Ability Question seemed to haunt me at every corner. Audience members would politely point out that one can know an explanation "without being able to use it in the right way," and understanding surely requires this kind of ability. Many argue that simply knowing an explanation does not involve the right kind of abilities to constitute understanding (De Regt 2004, 2009b; De Regt and Dieks 2005; Grimm 2010, 2014; Hills 2015; Newman 2012, 2013, 2014; Pritchard 2008, 2009a, 2010, 2014; Wilkenfeld 2013; Wilkenfeld and Hellmann 2014). With tongue planted firmly in cheek, let's call them "enablers." Virtually all enablers deploy some version of the following:

The Classic Ability Argument

CA1. Understanding-why requires some "special ability."
CA2. Knowledge-why does not require this ability.

CA3. ∴ It is possible to know-why without understanding-why; i.e., the received view is false. (CA1, CA2)

For instance, consider Pritchard's (2014, 316) counterexample to the received view:

> Kate comes to know that ... the introduction of the oxygen ... caused the chemical reaction not because she figured this out for herself, but because a fellow scientist, who has specialized expertise in this regard which our hero lacks, informs her that this is the cause of the reaction.[2]

Thus, Kate knows why the chemical reaction occurred. Yet, Pritchard goes on to deny her understanding:

> Crucially, however, Kate does not understand why the chemical reaction took place, because in order to possess understanding in such a case it is surely required that she should have a sound epistemic grip on why cause and effect are related in this way. Since Kate lacks this, she lacks understanding. One can thus have the relevant knowledge of causes (along with the relevant knowledge why) and yet lack understanding.

As the first premise of the Classic Ability Argument (CA1) stipulates, Pritchard posits that understanding requires a special ability: the "sound epistemic grip on why cause and effect are related." Having said that, different authors prize different special abilities. Other proposals include particular reasoning abilities; the ability to manipulate, construct, or use representations/models; a special facility with modal space; and the ability to extend or apply the explanation to new cases. I discuss many of these abilities in further detail later.

In the context of the Classic Ability Question, abilities are "special" just in case they are not entailed by knowledge-why. Seen in this light, the Classic Ability Question boils down to the veracity of CA1: enablers accept it; their critics deny it. When we turn to the Updated Ability Question, abilities are special if they could not be gained by bettering one's understanding in accordance with the EKS1's dictates.

Pritchard also appeals to the second premise of the Classic Ability Argument, CA2. However special abilities are construed, some kinds of knowledge frequently lack them – namely testimonial knowledge and memorial knowledge by rote. Hence, the examples used to motivate CA2 almost always use these "passive" forms of knowledge to make their point.

---

[2] In this passage, Pritchard's protagonist is actually named "Kate*" and is contrasted with a nearly identical epistemic agent "Kate." Since I won't be invoking this contrast, I have removed Kate*'s asterisk.

As we see in Pritchard's example, Kate gains her knowledge-why via testimony from a colleague.

3.2. Resisting the Classic Ability Argument

To be honest, I've never felt the pull of the Classic Ability Argument. It's always seemed to me that once you're armed with a correct explanation, you surely have *some* understanding of the explanandum regardless of any further upmarket abilities in the neighborhood. Furthermore, I've always thought that knowledge – and especially *scientific* knowledge – requires a fair amount of ability. The EKS Model allows me to add flesh to these skeletal impressions.

Specifically, I will argue that the EKS Model's account of minimal understanding has the resources to challenge the assumption that understanding requires special abilities (Sections 3.2.1–3.2.2). I will then argue that the EKS Model's comparative principles also can unseat the Classic Ability Argument's second premise by showing that scientific knowledge is replete with abilities (Section 3.2.3).

3.2.1. *Understanding Without Special Abilities*

A quick look at the Classic Ability Argument suggests a looming dialectical impasse – why would enablers' critics ever grant its first premise, CA1? To make things worse, enablers typically take this claim as obvious or intuitive but very rarely argue for it. As someone sympathetic to the received view, I don't share these intuitions. My goal in this section is not to persuade enablers. Rather, I want to sketch an alternative picture of understanding that is dominated by rather pedestrian abilities. In this way, the idea that understanding always requires some special ability is shown to be less than sacrosanct.

In one respect, the enablers' first premise is quite strong: it holds that understanding *always* involves special abilities. Anything weaker than this won't do the trick, for it would mean that understanding's affinity for ability is no deeper than knowledge's. But to appreciate how enablers overstate their case in this regard, we first need to remind ourselves that one person's understanding may be *better* or *worse* than another's. Enablers' intuitions seem most plausible when we ignore the lower end of this spectrum – when we overlook a kind of *minimal* understanding that might well bypass these special abilities. To that end, recall our definition of this concept from above:

(EKS2) *S* has minimal understanding of why *p* if and only if, for some *q*, *S* believes that *q explains why p*, and *q explains why p* is approximately true.

Since knowledge entails true belief, but not vice versa, minimal understanding involves no special abilities. Hence, if this account of minimal understanding is correct, then CA1 is false.

To get a better sense of EKS2, let's drill down into its core concept – an approximately true explanation. Recall my "theory" of explanation from Chapter 1:

q (correctly) explains why *p* if and only if:
(1) *p* is (approximately) true;
(2) *q* makes a difference to *p*;
(3) *q* satisfies your ontological requirements (so long as they are reasonable);[3] and
(4) *q* satisfies the appropriate local constraints.

This, of course, specifies the conditions for when "*q* explains why *p*" is *strictly* true. What about when it is only *approximately* so? Let's say that "*q* explains why *p*" is approximately true if and only if the first of these conditions is satisfied, and, furthermore, *some* of the terms in the explanans (*q*) that purport to make a difference to the explanandum (*p*) actually do make a difference and also satisfy your preferred ontological requirements.[4]

I assume that strictly true explanations are such that *all* of the putative difference-makers are genuine and ontologically licit. Additionally, I assume that only strictly true explanations must satisfy the more domain-specific local constraints, such as citing mechanisms, the use of certain idealizations, unification, and so forth.

So, to have minimal understanding is to have a good approximation of the explanandum and to have at least one ontologically admissible difference-maker in one's possession. Of course, I don't expect enablers to grant me this account of minimal understanding free of charge. Indeed, it's reasonable for anyone – enabler or otherwise – to ask why this meager epistemic state is any kind of understanding whatsoever.

---

[3] Recall that the caginess of this condition stems from a desire to sidestep debates about scientific realism. The general idea is that a theory of understanding can proceed while remaining neutral on this larger issue. See Chapter 6 for further details.

[4] Without a concrete ontological requirement, this gets a bit murky: do ontological requirements concern explanantia or the terms therein? I assume that any ontological requirement designed for one can be revised to accommodate the other.

Here's a quick and dirty argument: if one is able to explain something, then she has at least *some* understanding of why it is the case. But what does this so-called "ability to explain" entail? Not much: a fairly accurate representation of an explanation will do the trick. This needn't be anything more than what EKS2 stipulates: approximately true beliefs in correct explanations. Indeed, there might well be more primitive representational vehicles than beliefs that deliver this same result.[5]

Recall that a special ability is something not required by knowledge-why (i.e., by *explanatory* knowledge). Thus, the ability to explain is *especially unspecial*. In other words, there appears to be a direct route from explanations and understanding-why that makes no detour through special abilities. As an illustration, consider a variation on Pritchard's example:

Anne and Bob

Suppose that two people wonder why a particular chemical reaction occurred. The first person, Anne, can correctly answer this question: because oxygen was introduced. The second person, Bob, stares back blankly, with no answer to speak of.

Here is my intuition: Anne understands why the chemical reaction occurred better than Bob does. But, of course, this entails that Anne understands *to some degree* – that she has achieved *minimal* understanding at the very least. Moreover, to make this comparison between Anne and Bob, we needn't know anything about Anne's special abilities. We needn't know whether she has the linguistic capacities to assert "Because of the oxygen" when asked "Why did the reaction occur?"; whether she can put this information to use in novel contexts; whether she can construct fancy models; and so on. Indeed, we also needn't know whether she arrived at that answer through some rich, luck-free etiology, testimony, rote memorization, or even if she has any justification for her answer.[6] Simply having the correct answer is enough. And, of course, if Anne has minimal understanding, then so does the protagonist of Pritchard's original example, Kate. The key is recognizing that Kate and Anne only exhibit a *modest* amount of understanding. Those with special abilities might well exhibit *better* understanding, but that's beside the point (at least for now).

---

[5] While I would be happy to have an even more austere account of minimal understanding, I won't pursue this here. Newman (2012, 2013) and Wilkenfeld (2013) are two modest enablers who, with a bit of finessing, might lend themselves to this kind of account.

[6] Here, I'm loosely borrowing from a nice argument in Morris (2012), who shows some (but in my estimate, insufficient) appreciation for degrees of understanding.

In short, minimal understanding appears to require something *less* than knowledge of an explanation – never mind some special ability that requires *more* than knowledge. As I said, I don't expect this to persuade enablers. However, I do think it belies the theory-neutrality of CA1. That claim should not be assumed; it must be earned through argument.

3.2.2. *Minimal Understanding Defended*

To continue to chip away at the intuition behind CA1, let me anticipate and rebut two objections to my account of minimal understanding. The first is what I call the *Parroting Objection.*[7] According to the preceding discussion, Anne's approximately true belief in an explanation affords her (minimal) understanding. However, when we think of how this understanding can be used in other contexts, suspicions arise. For instance, teachers frequently use why-questions to test whether their students have a basic "understanding" of a given subject matter. However, the objection comes to roost in precisely these sorts of contexts, for it seems that students frequently "parrot" information without really understanding it. More precisely:

The Parroting Objection

PO1. If minimal understanding is having an approximately true belief in an explanation, then someone's merely reciting (parroting) a correct explanation affords them understanding of the explanandum.

PO2. Merely reciting a correct explanation does not afford understanding of an explanandum.

PO3. ∴ Minimal understanding requires more than an approximately true belief in an explanation; i.e., EKS2 is false. (PO1, PO2)

I shall argue that the first premise is false. Specifically, beliefs require concept-possession, and the kinds of mindless recitation in question do not.

Begin with the claim that beliefs require concept-possession. If Kate believes that *the chemical reaction occurred because oxygen was introduced,* then she must correctly apply the concepts of *chemical reaction, because, oxygen,* and so on. Had she lacked any of these concepts, then her belief could (at best) only be some coarser-grained neighbor. For instance, absent certain chemical concepts, the content of her belief might be that *that thing in the lab happened because that stuff over there was introduced.*[8]

---

[7] Thanks to Emily Sullivan for pushing me on this point.

[8] There may be externalist views about mental content that would conflict with this part of my rebuttal. However, these views will put more pressure on PO2.

Turn next to the claim that mere recitation is devoid of this kind of concept-possession. This is easiest to see with real parrots, who regularly squawk well-formed syntax with nary a concept in sight. Hence, more often than not, they cannot believe what they squawk. So far as I can tell, people who parrot explanations are no different than their feathered friends. They, too, cannot (or at least do not) believe what they squawk. But since my account of minimal understanding requires belief, it does not engender crude parroting. Hence, absent further clarifications about which abilities are absent in cases of brute recitation, the objection misses its mark.

We might see the second objection as filling in some of these abilities that are on holiday in cases of parroting. Following Woodward (2003), some authors hold that a special ability associated with understanding involves the anticipation of what would happen to a system if it were to change in various ways (Grimm 2014; Hills 2015). Specifically, Woodward argues that an explanation should be able to answer "what-if-things-had-been-different" questions ("what-if" questions for short). To see the importance of what-if questions, let's return to Pritchard's example. Let Molly be the expert colleague who told Kate about the oxygen. Then we might expect Molly to have true beliefs such as the following:

(1) The chemical reaction released lithium oxide (rather than **lithium hydride**) because oxygen (rather than **hydrogen**) was introduced.

In other words, Molly can correctly infer what would have happened if something had been different; namely had hydrogen rather than oxygen been introduced. For this reason, Molly understands why the chemical reaction occurred. Since Anne and Kate lack this ability to answer what-if questions, proponents of this view might resist my account of minimal understanding. More precisely:

The What-If Objection

WI1. To understand, one must be able to perform what-if reasoning.

WI2. If minimal understanding is having an approximately true belief in an explanation, then someone can understand without being able to perform what-if reasoning.

WI3. ∴ Minimal understanding requires more than an approximately true belief in an explanation; i.e., EKS2 is false (WI1, WI2)

While the example of Molly illustrates the motivations behind the first premise, the other premise, WI2, denies that minimal understanding requires what-if reasoning. However, I see no reason to grant this premise. Indeed, my reply to the What-If Objection is simply an elaboration of my reply to the Parroting Objection. Specifically, belief in an explanation requires that one possess the concept of *explains* or *because*, and possession of this concept requires that one be able to answer what-if questions.[9]

Begin with my first claim, that believing an explanation requires correct application of the concept of *explains* or *because*. If this were not so, one would fail to have an answer to a why-question *qua* answer. For instance, suppose that Kate had true beliefs that a chemical reaction occurred and that oxygen was introduced, but failed to see that the former was *because* of the latter. Then she clearly does not *believe* that the introduction of oxygen explains why the reaction occurred.

Turn now to my second claim, that possessing the *explains*-concept requires that one can answer what-if questions. Like understanding, concept-possession admits of degrees. To possess a concept is to be able to use it correctly. But most concepts have a multitude of correct uses. A natural proposal is that one person has greater mastery of a concept if the former can use that concept in more correct ways than the latter; e.g., mastering the different roles it can play in different inferences (Brandom 1994; Sellars 1963).[10]

But *which* inferences? Recall that, on my view, an explanation is correct only if its explanans *makes a difference* to its explanandum and that difference-making is naturally glossed in terms of counterfactual dependence (see Section 1.3.1). Thus, the inferences one must be able to make in order to wield the *explains*-concept are precisely the counterfactual or what-if inferences that are in question. If someone lacked the ability to reason in this manner, it becomes doubtful that the person is representing an *explanatory* statement. For instance, if an agent has no sense of how the explanans and explanandum are connected, she may have nothing more than a belief in a *conjunction*, but not a belief in an *explanation*. If she recognizes some kind of connection that falls short of the (modest) counterfactual dependence that I require, then she might have a belief in an *inference*, but not in an *explanation*. So, the ability to reason counterfactually is *essential* to minimal understanding. Thus, although minimal understanding clearly involves no *special* abilities, it does involve *some*

---

[9] I'm also happy to grant that inquirers can sometimes achieve understanding when they possess only "thicker" explanatory concepts, e.g., *causes, unifies*. I bracket this hereafter.

[10] While I use Brandom's account of concept possession below, my arguments would be unaffected by an alternative account, so long as it entails that concept mastery increases in proportion to the number of ways one can use a concept.

abilities. I suspect that this point has been overlooked in the Classic Ability Argument.

In the context of minimal understanding, the capacity for answering what-if questions needn't be too robust. Specifically, minimal understanding might only require the ability to infer that the explanation can be used to answer *some* what-if questions while being unable to identify *which* of these questions it can answer and also being unable to answer these questions in much detail.

More concretely, assume that Kate, the recipient of Molly's testimony, has minimal understanding. Her true belief might consist only in the following:

> (2) This (rather than **some other**) chemical reaction occurred because oxygen (rather than **some other element**) was introduced.

While this is considerably less impressive than Molly's what-if reasoning, it is something. Indeed, since Kate has the concepts of *oxygen, because, chemical reaction*, and the like, she has minimal understanding on my view.

Importantly, in virtue of her possession of the *because*-concept, Kate can draw the following counterfactual claim from her belief in (2):

> (3) Had some other element been introduced, then some other chemical reaction would have occurred.

In other words, although her answer is modest, Kate can correctly answer the what-if question; namely, "What if another element had been introduced?" with "Another chemical reaction would have occurred." These points generalize to all of minimal understanding.[11]

Summarizing, I think it is possible to have minimal understanding without any special abilities. Furthermore, I have been using *minimal* understanding as a limiting case. Many instances of low-grade but non-minimal outright understanding will yield the same result. Indeed, I have only had to invoke *approximately true beliefs* to make my case, but presumably we can extend that argument using the more demanding epistemic status of *knowledge*. In short, there are plenty of situations where we understand without any special ability.

---

[11] Extrapolating, minimal understanding must yield beliefs in propositions of the following form:

$$\exists i(Y = y(rather \text{ than } y_i) \text{ } because \text{ } X = x(rather \text{ than} x_i)).$$

Here, capital letters denote variables; lowercase letters, their respective values; $x \neq x_i$; $y \neq y_i$. One will then be able to answer the question, "What if X had been different?" with the modest answer, "Then Y would also have been different."

3.2.3.   Scientific Knowledge and Ability

Given the difficulties that minimal understanding poses to the Classic Ability Argument, it's fair to ask whether enablers are interested only in more demanding kinds of understanding. Most theorists of understanding appear concerned with *outright* understanding (i.e., whether or not a person understands "full stop"). Furthermore, enablers have a particular set of abilities in mind when they reach this stop. Could this be where the true force of the Classic Ability Argument lies?

To get our bearings, recall that I offered a contextualist semantics for outright understanding in Chapter 1:

> "*S* understands why *p*" is true in context *C* if and only if *S* has minimal understanding and *S* approximates ideal understanding of why *p* closely enough in *C*.

This would suggest that we revise the Classic Ability Argument as follows:

> CA1*. Some instances of outright understanding require a special ability.

Here "outright understanding" is construed in theory-neutral terms; i.e., not as the EKS Model ultimately defines it in terms of scientific knowledge and explanation, but in some pre-theoretic or intuitive sense.[12]

While the proposed revision, CA1*, would block the arguments of the previous two sections, it faces its own challenges. The EKS Model can be used to critique the other premise of the Classic Ability Argument, CA2, which states that knowledge does not entail certain understanding-relevant abilities. Specifically, when we switch from the received view to the EKS Model, this premise must be revised accordingly:

> CA2*. Even the most demanding instances of *scientific* knowledge-why do not require this special ability.[13]

We can now see why CA2* spells trouble for the Classic Ability Argument. Suppose that scientific knowledge requires some ability. Then, per CA2*, this ability cannot be special. Since scientific knowledge has more demanding ability-requirements than non-scientific knowledge,

---

[12] Note that the preceding definition of outright understanding might still be theory-neutral so long as "minimal understanding" and "better understanding" are also theory-neutral; e.g., not in terms of the EKS Model or any other specific account of understanding.

[13] Dialectically speaking, I *could* compare enablers' proposals with *ideal* understanding. However, my arguments won't appeal to the ideal. They will appeal to abilities that scientists exercise in a typical workday.

this contracts the space of special abilities significantly. In other words, CA2* becomes far less plausible.

To cast a shadow on CA2*'s prospects, I conclude this section by canvassing the abilities characteristic of scientific knowledge of an explanation. Earlier, I argued that scientific explanatory evaluation (SEEing) characterizes this kind of knowledge. Let's remind ourselves of SEEing's three core features. First, scientists must *consider* plausible potential explanations of the phenomenon of interest. Inquirers may consider explanations either by generating new hypotheses from scratch, or (more commonly) they may countenance explanations that have been generated by others.

Second, the potential explanations that have been considered must be *compared*. Here, scientists determine the relationships between the explanations by drawing out the consequences of the different explanations and gathering scientific evidence to test those consequences. Other explanation-improving factors (such as simplicity, scope, fruitfulness, fit with background beliefs) may also be germane in this stage. Comparisons indicate whether the different explanations compete or complement each other, which is best, how different explanations interact, and the like. In paradigmatic cases, one explanation is the "winner" of these comparisons.

Third, scientists *form* their doxastic attitudes on the basis of these comparisons. Scientists believe that clear winners in the prior stage of comparison are true, disbelieve clear losers, and assign appropriate degrees of belief about the middle of the pack.

At this point, enablers may deny that these abilities are actually *part* of scientific knowledge and may insist that they are simply *independent addenda* to that knowledge. If this objection were sound, SEEing-related abilities would simply be special abilities.

However, the objection is flawed. As Pritchard (2012, 248) argues elsewhere, our concept of knowledge is informed by the "ability intuition"; i.e., the idea that "knowledge requires cognitive ability, in the sense that when one knows one's cognitive success should be the product of one's cognitive ability."[14] For instance, perceptual knowledge is the product of perceptual abilities; inferential knowledge, the product of inferential abilities; memorial knowledge, the product of memory; and so forth. By parity of reasoning, explanatory knowledge should be the result of "explanatory abilities." SEEing nicely describes some core features of these explanatory abilities. I defend these claims at greater length in Chapters 7 and 8.

---

[14] As Pritchard acknowledges, the ability intuition has received its most rigorous development from virtue epistemologists (e.g., Code 1987; Greco 2009; Sosa 1980, 2007, 2009; Zagzebski 1996).

For now, note that each aspect of SEEing involves significant cognitive abilities. For instance, consideration involves highly structured creativity (when generating alternative explanations) and mastery of background literature (when countenancing extant alternatives). Comparison involves insight into different explanatory relationships (e.g., causal structures, dependency relationships, inferential connections within and between explanations), the ability to draw out predictive consequences of each explanation, and various kinds of methodological prowess, such as the ability to design experiments and interpret results. Formation deploys inferential abilities.

These abilities have a lot going for them. For instance, they readily explain why testimony and rote memorization are unlikely to yield much understanding: these do not require much in the way of SEEing.[15] Consequently, if Kate is being held to higher standards than those suggested in Section 3.2.1 and Section 3.2.2, the EKS Model can readily explain why she lacks understanding. Nevertheless, in Sections 3.4 to 3.6, I want to be dialectically opportunistic: offer me up a special ability, and I'll show you that it either plays a role in SEEing or leads to counterintuitive results about understanding.

3.3. The Updated Ability Argument

To some extent, the preceding responses to the Classic Ability Argument pull us in opposite directions. On the one hand, denying the first premise rested on the idea that understanding required no abilities over and above those involved in having a true belief about an explanation. On the other hand, denying the second premise appealed to the idea that understanding is a kind of *scientific* knowledge, which requires significant abilities.

To navigate this tension, we need to account for degrees of understanding. Once we grant that *low-grade* understanding is devoid of special abilities, how do we go about giving abilities some role in *high-grade* understanding? This, of course, is simply to ask the Updated Ability Question:

---

[15] It's uncontroversial that scientists *sometimes* acquire knowledge of explanations through testimony. However, that won't be scientific knowledge in my intended sense, as described in Chapter 1. On my view, scientists who gain explanatory knowledge via testimony still have *some* degree of understanding, but it is considerably less than a scientist who came to know that same explanation via SEEing.

- Which abilities, if any, improve our understanding?

The EKS Model provides a straightforward answer to the Updated Ability Question: only the abilities that figure in scientific knowledge of an explanation are the ones that improve our understanding. The Scientific Knowledge Principle is the key:

> *Ceteris paribus*, if S_1's grasp of p's explanatory nexus bears greater resemblance to scientific knowledge than S_2's, then S_1 understands why p better than S_2.

Our grasp is firmer or looser depending on how closely it approximates the abilities that furnish scientific knowledge of an explanation. Here's a toy example: if an explanation is scientifically confirmed by way of two experiments, and you know how both of those experiments support the explanatory hypothesis, but I only know how one of them supports this hypothesis, then, *ceteris paribus*, your understanding is better than mine. Of course, scientific knowledge of an explanation can be more closely approximated in many other ways.

However, just as we should update the received view by way of the EKS Model, we should assess how the latter fares against a revised version of the Classic Ability Argument:

Updated Ability Argument

UA1. S_1 and S_2 are otherwise identical with respect to their grasp of why p.
UA2. S_1 has a special ability and S_2 does not.
UA3. If UA1 and UA2 are true, then S_1 understands why p better than S_2.
UA4. If UA1 and UA2 are true, then, according to the EKS Model, S_1 does not *understand why p better than S_2*.
UA5. ∴ The EKS Model is false. (UA1–UA4)

Whereas the Classic Ability Argument indicts knowledge-driven accounts of understanding for giving too much credit to unskilled understanders, the Updated Ability Argument indicts the EKS Model for undervaluing a special ability's capacity to improve our understanding in extra-scientific ways. Thus, like proponents of the EKS Model, new-wave enablers can grant that Kate has *some* understanding but still insist that knowledge-driven accounts of understanding fail to explain why her colleague has a *better* understanding of why the reaction occurred.

Despite these initial attractions of the Updated Ability Argument, it faces a formidable dilemma. Enablers can either grant or resist the idea that scientific knowledge-why entails their favorite special abilities. If they grant this entailment, then special abilities don't look all that special, and there's no reason for enablers to deny the EKS Model.[16] Specifically, they will have conceded that UA4 is false. In this case, let's say that a special ability is *redundant* given the EKS Model. By contrast, if enablers deny that scientific knowledge entails their special abilities, then there are good reasons to think that the abilities will be erroneous to understanding. In this case, UA3 will be false (or at least unsubstantiated), making the special ability *excessive*. Hence, special abilities are either redundant or excessive. Either horn undermines a premise in the Updated Ability Argument, so it cannot touch the EKS Model. In what follows, I inflict this dilemma on three leading enablers: Pritchard (Section 3.4), Hills (Section 3.5), and Grimm (Section 3.6).[17]

3.4. Pritchard on Cognitive Achievements

Pritchard's account of abilities is richer than the preceding discussion might suggest. His work consistently emphasizes that understanding-why is a "cognitive achievement." A cognitive achievement is a cognitive success because of cognitive ability. In the context of understanding, a cognitive success is a true belief in an explanation. To be an achievement, such a belief must result from exercising a significant skill or overcoming a significant obstacle.

While the preceding arguments (in Section 3.2) belie this tight connection between cognitive achievement and *minimal* understanding,[18] Pritchard's view may still make good sense of more *demanding* kinds of understanding. In what follows, I will consider two variations on Pritchard's ideas. The first is that the *only* defining feature of high-grade understanding is that *some* kind of cognitive achievement results in a correct explanation. The second is that high-grade understanding is a

---

[16] At least, there's no reason for enablers to challenge the EKS Model *qua* enablers. Both Pritchard and Hills deny that understanding has the same relationship to luck as knowledge. I address this objection in Chapter 7.

[17] I have critiqued other enablers elsewhere. For instance, De Regt (2009a, 2009b) and De Regt and Dieks (2005) are subjected to something akin to the aforementioned dilemma in Khalifa (2012). Some of those critiques are restated in Section 2.5. In a thoughtful discussion of that work, Newman (2014) – also an enabler – offers rebuttals to these critiques and suggests that his view (Newman 2012, 2013) provides some stiffer challenges. I respond to these claims in Khalifa (2015). While there have been some small changes in my view since then, my ideas about abilities are consistent with the position advanced in this book.

[18] Carter and Gordon (2014) provide a lengthier discussion of this lacuna in Pritchard's account.

more *specific* kind of cognitive achievement. I will argue that both variations succumb to the dilemma of redundancy and excess.

First, let's suppose that Pritchard does not intend any particular abilities to be special but is content to have *any* ability that figures in a cognitive achievement resulting in a true belief about an explanation as both necessary and sufficient for understanding; i.e.,

> S understands why p if only if:
> (A) For some q, S has the true belief that q *explains why* p;
> (B) This belief is because of one of S's cognitive abilities; and
> (C) The deployment of this cognitive ability involves exercising a significant skill or overcoming a significant obstacle.

Let's grant that this will allow Pritchard to satisfy condition UA4 of the Updated Ability Argument; i.e., that the EKS Model cannot capture all of the ways in which understanding improves on this "broad cognitive achievements model" of understanding. The latter model will still fail with respect to UA3: it will be excessive. To see this, imagine that one of Kate's coworkers, Nate, has both exceptional vision and world-class lip-reading skills. From a remarkable distance, he learns the cause of the chemical reaction by eavesdropping on Kate's conversation with Molly. While I have no doubt that this is a remarkable cognitive achievement, Nate does not understand why the chemical reaction occurred any better than does Kate. Consequently, I assume that Pritchard does not propose such a capacious account of special abilities.

Unfortunately, Pritchard is not altogether clear about how he would constrain the kinds of abilities that can figure in understanding. One of his more suggestive proposals is that the ability must involve "some sort of grip on how this cause generated this effect, a grip of the kind that could be offered as an explanation were someone to ask why the event occurred" (Pritchard 2014, 321). However, this conflates two different explanations: (1) an explanation of how, e.g., oxygen caused the reaction versus (2) an explanation of why the reaction occurred at all. For instance, since Kate knows why the reaction occurred, she can "offer an explanation were someone to ask why" this is so, even if she cannot explain how oxygen caused the reaction. Hence, on the second interpretation, simple knowledge-why affords one the abilities that Pritchard claims are special. However, given that special abilities are precisely those not entailed by merely knowing-why, this proposal fails to debunk the received view – much less the EKS Model.

So, something in the vicinity of the first interpretation must be what is intended. Specifically, combining this with the idea that understanding is a cognitive achievement suggests the following:

> S understands why p if and only if:
> (A) For some q, S has the true belief that q *explains why* p;
> (B) This belief is because of S's "grip" of how q explains why p; and
> (C) This "gripping" involves exercising a significant skill or overcoming a significant obstacle.

On this view, understanding is what I will call a "why-how achievement:" true beliefs about *why* something is the case are cognitive successes and are the result of one's grip or ability to follow *how* an explanation hangs together.

As I'll now argue, this proposal faces three problems. First, it raises analogous worries to those in Section 3.2. There, I argued that some (low-grade) understanding could exist *sans* ability. Pritchard anticipates this objection to the Classic Ability Argument, and invokes understanding as a why-how achievement in his rebuttal:

> ... there is a distinction to be drawn between, on the one hand, having a sufficient conception of how cause and effect might be related to enable the agent to have the relevant causal knowledge, and, on the other hand, having a sufficient explanatory grip on how this particular cause generated this particular effect in order to possess the corresponding understanding. (Pritchard 2014, 322)

His point here is that this distinction is not merely a difference in degree, but a difference in kind. However, Pritchard has missed the relevant distinction. According to Pritchard, Kate has the following true belief:

> (4) The introduction of the oxygen caused the chemical reaction to occur.

Given the way that she acquired this belief, she *knows* why the reaction occurred. According to Pritchard, in order to *understand* why the reaction occurred, Kate must have a further grip on how the oxygen caused the reaction; e.g., a true belief that:

> (5) The introduction of the oxygen caused the chemical reaction to occur because the oxygen interacted with lithium.

However, it's quite clear that both (4) and (5) are explanations. In particular, the first answers the question, "Why did the chemical reaction occur?" and the second, "How did the introduction of oxygen cause the chemical reaction to occur?" So, the difference is only one of degree: a person who has grasped (5) has more explanatory information than a person who has only grasped (4). But this readily fits within the EKS Model, for the Nexus Principle entails that the former person's understanding is better:

> Ceteris paribus, if S_1 grasps p's explanatory nexus more completely than S_2, then S_1 understands why p better than S_2.

Consequently, any "grasp" or "grip" of (5) isn't special. Because the Nexus Principle is clearly part of the EKS Model, Pritchard's proposal is redundant.

Now, perhaps the Nexus Principle's appeal to "grasping" obscures this point, as grasping seems to involve some special ability. However, at least as far as Pritchard is concerned, we've been given no reason to think that even the *received view* fails to provide sufficient conditions on understanding. For all we've been told, *knowledge* of (5) provides the "grip" that Pritchard claimed as a special ability. Moreover, the Scientific Knowledge Principle is more flexible than the received view on this front since knowledge is only one of many ways to approximate scientific knowledge.

Perhaps Pritchard's more ambitious ideas about cognitive achievements might shed some further light on the nature of "grasping" or "gripping." This leads to my second objection: his specific achievement-driven machinery is unnecessary. Many cognitive achievements improve one's understanding yet lack Pritchard's "gripping." For instance, consider the following:

> Through careful experiments, Molly eliminates nitrogen, carbon dioxide, etc. as explanations of the chemical reaction, isolates oxygen as the main reactant, and correctly infers that oxygen explains why the reaction occurred. However, Molly has no belief as to how oxygen caused the reaction.

Clearly, Molly has a true belief about an explanation that is because of exceptional skills – namely those involved in SEEing. Hence, she has a cognitive achievement. It also seems uncontroversial that she understands why the chemical reaction occurs, yet she has nothing resembling Pritchard's grip. Stated more generally, arriving at the correct explanation q of p by SEEing involves a variety of abilities. Yet these abilities needn't include the ability to answer the question, "How does q cause/explain why p?" Indeed, one can have no great answer to this question, while still having a solid answer to the question "Why p?"

Furthermore, Pritchard's account gets the majority of scientific practice backward. Scientists tend to *first* discover that A causes B, and, on the basis of that discovery, they *subsequently* understand the mechanisms that intervene between A and B. For instance, scientists first discovered that smoking causes lung cancer, and, because of that discovery, they subsequently discovered that this was due to tar in the lungs. In other words, "how-why achievements" are far more prevalent than Pritchard's "why-how achievements." Consequently, Pritchard's account fails to identify a large swath of advances in scientific understanding.

Thus, the second problem is that science does not rely exclusively on why-how achievements to garner understanding. Indeed, scientists are more likely to gain true answers to how-questions because of their ability to answer why-questions. Pritchard's account requires precisely the opposite. By appealing to scientific practice directly, the Scientific Knowledge Principle avoids this problem. While this does not bear directly on the Updated Ability Argument, it's a nice bonus.

Here's my third and final worry: absent a clearer account of what's involved in "gripping," it's very hard to see how one's true beliefs about an explanation could be *because* of this ability. That is, if one comes to a belief because of an achievement, the belief-forming process involved in that achievement must be reliable. However, when isolated from SEEing, Pritchard's cited ability is unreliable. Consider the following:

> Through sheer speculation, Dave posits a Rube-Goldberg-like mechanism linking a cause q to its effect p and comes to believe that q causes p because he is so convinced by his fanciful mechanism. The Rube-Goldberg-like mechanism is ingenious in its hypothetical design, requiring incredible insight, creativity, and theoretical acumen to even imagine, but Dave has absolutely zero empirical evidence that this elaborate mechanism connects q to p. As it turns out, q actually causes p, and, by dumb luck, Dave's speculation about the mechanism turns out to be correct.

As Dave's lucky speculation makes clear, unless Pritchard clarifies what he means by "gripping," the why-how reasoning that he champions is not a cognitive achievement since the true belief in an explanation is not *because* of one's grip of how the explanans relates to the explanandum. This suggests a further problem in pitting Pritchard's why-how achievements against the EKS Model; it's unclear that why-how reasoning is robust enough to explain the truth of one's explanatory commitments.[19] Importantly, Dave's "grip" suffers primarily because it lacks any empirical

---

[19] More precisely, on Pritchard's version of the Updated Ability Argument, it's hard to see how why-how reasoning can establish UA2. As the example makes clear, why-how reasoners may be identical to other agents whose true beliefs are not because of their abilities.

support. Since such confirmation lies at the heart of scientific knowledge, the EKS Model remains the more attractive option.

Thus, Pritchard's account of understanding as a why-how achievement does not pose a serious threat to the EKS Model. Indeed, by embracing the EKS Model, his account would be spared from all three of the difficulties just canvassed. For all that Pritchard has shown, understanding is a cognitive achievement, but only insofar as it respects the good sense of scientific practice.

3.5.   Hills on Cognitive Control

Hills (2009, 2015) enables understanding in a different way than Pritchard, though she also offers a version of the Classic Ability Argument. My criticisms in Section 3.2 apply just as well to Hills. However, it's an interesting question as to how Hills's understanding-mongering abilities figure in the Updated Ability Argument. Standing at the core of Hills' view is that understanding is a certain kind of know-how or ability. More precisely, for Hills,

> if you understand why p (and q is why p), then you believe that p and that q is why p and in the right sort of circumstances you can successfully:
> (a) follow some explanation of why p given by someone else.
> (b) explain why p in your own words.
> (c) draw the conclusion that p (or that probably p) from the information that q.
> (d) draw the conclusion that p' (or that probably p') from the information that q' (where p' and q' are similar to but not identical to p and q).
> (e) given the information that p, give the right explanation, q.
> (f) given the information that p', give the right explanation, q'.

Hills calls this set of abilities "cognitive control" over p and q. In what follows, I discuss each of Hills' conditions, arguing that each is either excessive or redundant.

First, let's consider the "linguistic abilities," (a) and (b). I will say more about these in Chapter 5, but the basic idea is this: either explanations can be tacit or they cannot. If explanations can be tacit, then the link between language and explanation is weak. Consequently, (a) and (b) would be excessive: following others' explanations and putting explanations in your own words would not improve understanding-why. Alternatively, if explanations cannot be tacit, then understanders with explanations have greater linguistic capacities. In this case, Hills' conditions (a) and (b) are redundant, for they would replicate the ways in which scientists represent the explanatory nexus. Hence, on either account of the relationship between language and explanation, Hills cannot free (a) and (b) from the dilemma of redundancy and excess.

For reasons discussed in the context of De Regt's view – what I called "the problem of improbable explananda" (Section 2.5.4) – conditions (c) and (d) are clearly excessive. Simply put, some explanations are not inferences. The classic example, owing to Scriven (1959), is that untreated syphilis explains why one has paresis, yet only a small percentage of untreated syphilitics contract paresis. For any explanations with a similar, non-inferential structure, this means that (c) and (d) cannot be jointly satisfied with (e) and (f). Hence, these sorts of examples will automatically falsify Hills's view. By contrast, the EKS Model puts few *a priori* constraints of this sort on explanations: they are largely as we find them in the sciences.[20]

Condition (e) – that one be able to explain what one understands – will clearly be redundant given the EKS Model. Indeed, Hills simply notes that one must be able to exercise this ability in the "right sort of circumstances." By contrast, my account of SEEing opens up this black box in her account: the "right circumstances" arise under careful consideration and comparison of plausible alternative explanations.

Condition (f) – that one be able to explain a similar phenomenon – is either redundant or excessive. On the first horn, scientists can frequently explain similar cases. When the condition fails to follow the EKS Model, it falls on the second horn and becomes excessive. For instance, suppose that I can give the following correct explanation:

(6) There are fewer cookies in the cupboard because John ate some.

This explanation resembles the following:

(7) There are fewer brown anoles on some Bahamian islands because northern curly-tailed lizards ate some.[21]

However, understanding why there are fewer cookies in the cupboard surely does not improve when I can explain why there are fewer brown anoles on some Bahamian islands. If someone possesses the explanations about the cookies (6) and about the brown anoles (7), she understands why there are fewer cookies in the cupboard and she *also* understands why there

---

[20] For a view of the constraints that I do put on explanations, consult Chapter 1.

[21] This is a slight variation on Schoener, Spiller, and Losos (2001); more on brown anoles in the next chapter.

are fewer brown anoles on some Bahamian islands. So, she understands more things. However, it does not follow from this that she understands those things better.

Now perhaps Hills intends that the explanations bear stronger resemblance to each other than cookies and anoles. For instance, perhaps understanding why there are fewer cookies requires the ability to give the following explanation:

(8) There are fewer crackers in the cupboard because Jane ate some.

Yet this example of similar explanations suffers the same fate as (7). Just as with anoles, a person may understand why there are fewer cookies (6) and also understand why there are fewer crackers in the cupboard (8), without shoehorning these two explanations into a single instance of understanding. Hence, even in highly similar cases, (f) may well be excessive; i.e., there is no reason to grant that the capacity to give explanations of similar phenomena enhances understanding.

A much more plausible alternative to Hills's final condition is suggested by the methodology of controlled experiments. Ideally, the control and experimental groups are identical save for two differences: one in the effect of interest and another in the cause of that effect. However, frequently this means that a proposition of the form *not-p* best describes the control group and *p* best describes the experimental group. *Pace* Hills, it is hard to imagine two less "similar" propositions. Similarly, in the context of controlled experiments, if *q* explains why *p*, then *not-q* frequently explains why *not-p*. By contrast, this kind of reasoning is very close to that required by the account of minimal understanding presented in Section 3.2, and that improves in proportion to one's understanding. Hence, the EKS Model appears to offer a more plausible account of the explanations one must possess in order to understand.

3.6.   Grimm on Grasping

Thus far, I've defended the EKS Model from Pritchard's and Hills's alternatives. Unlike Pritchard and Hills, Grimm embraces the idea that understanding is a species of knowledge. Indeed, while Grimm equates his account of understanding with "knowledge of causes," his more considered view follows Jaegwon Kim's (1994) idea that understanding is knowledge of a wider variety of "dependency relations," the common core of which is their ability to underwrite explanations. Hence, Grimm and I both take understanding to be a kind of explanatory knowledge.[22]

For this reason, Grimm's work shows a special sensitivity to the received view and its attendant challenges. Throughout his writings, we can see three different responses to the Classic Ability Argument. First, he offers arguments very similar to those in Section 3.2 (Grimm 2014, 337–338). Since I have expanded on these arguments substantially, I will say nothing further about this strand in Grimm's thought. The second response invokes the ability to reason counterfactually. This ability is central to understanding – but only insofar as it is situated within the context of SEEing (Section 3.6.1). Finally, Grimm sometimes adverts to a kind of "modal apprehension" that is inspired by the rationalist epistemology of Bonjour (1998). I will argue that this is excessive (Section 3.6.2).

3.6.1. *Counterfactual Reasoning*

As just noted, Grimm spends some time defending the received view. Recently, however, he has devoted far more effort to distancing himself from that position. According to Grimm (2014, 336), we should contrast:

(a) assenting to a causal proposition on reliable grounds (on the basis, say, of reliable testimony, or reliable memory)
(b) seeing or grasping the modal relatedness of the terms of the causal relata.

Intuitively, the received view only gives us the first, (a). By contrast, Grimm takes the second, (b), to be more characteristic of understanding. Grimm sometimes construes this "seeing" or "grasping" as "the ability to anticipate how changes in the value of one of the variables . . . would lead to (*ceteris paribus*) a change in the value of another variable," plus the ability to apply general expressions of these dependency relations between variables to particular cases (2010, 340–341). Call this "counterfactual reasoning."

For example, suppose that someone recognizes (via Bernoulli's principle) that the shape of an airplane's wing (curved on top and flat on the bottom) creates a difference in the velocity of air on the top and the bottom of the wing, such that the pressure exerted by the slower moving air along the bottom of the wing is greater than the pressure exerted along the top of the wing. As a result of this difference in pressure, flight is possible.

---

[22] Parallel points apply to Greco (2013). For reasons discussed throughout the book, I construe explanation in less metaphysically laden terms than Greco, Grimm, or Kim. Nevertheless, this amounts to family infighting in the current context.

Moreover, assume that the person can see that changes in the shape of the wings or in the pattern of airflow would result in the plane not being able to fly. Intuitively, this person understands why planes fly, just as Grimm's account of counterfactual reasoning proposes.

Is counterfactual reasoning a special ability? Grimm certainly seems to suggest this at times. For instance, it plays some role in his abandoning the received view – what he calls the "propositional model" of understanding as "knowledge of causes." In its place, he claims that understanding ought to be understood as *non-propositional* knowledge of causes. On this view, counterfactual-reasoning abilities replace belief.

This has some plausibility. After all, unlike beliefs, abilities needn't be propositional attitudes. That being said, we have already recovered precisely the same counterfactual-reasoning ability in our discussion of minimal understanding: to have the *explains*-concept at all is to have the ability to answer what-if questions. For instance, recall our earlier example in which Molly knows that the chemical reaction produced lithium oxide (rather than lithium hydride) because the reactant was oxygen (rather than hydrogen). In this case, Molly is able to answer a what-if question, namely "What if hydrogen rather than oxygen were the reactant?," and hence anticipate changes in variables in precisely the way that Grimm suggests. Such an example is consistent with the EKS Model and delivers precisely the kind of counterfactual reasoning ability that Grimm seeks.

Generalizing, Section 3.2.2 suggests that, according to the EKS Model, many cases of understanding require S to know that $Y = y$ (rather than y') because $X = x$ (rather than x'). Wherever this is so, Grimm's counterfactual reasoning requirement is redundant. However, this way of collapsing Grimm's account into my own can be developed further. Specifically, counterfactual-reasoning abilities divorced from SEEing become excessive. For instance, consider the following consequences of the explanation of the chemical reaction:

(9)
(a) Had oxygen never ever existed, then the reaction would not have occurred.
(b) Had we called oxygen "dephlogisticated air," the chemical reaction still would have occurred.

If a person could only provide these kinds of answers to what-if questions, then her understanding seems quite modest. Contrast that with more informative kinds of counterfactual conclusions:

(10)
(a) Had the reactant been hydrogen (rather than oxygen), then the chemical reaction would have produced lithium hydride (rather than lithium oxide).
(b) Had there been no lithium in the test tube, then the chemical reaction would not have occurred.

Why does knowledge of the latter two counterfactuals provide greater understanding than their counterparts? A natural suggestion is that they are the kinds of things that could play a role in SEEing. Whereas oxygen never ever existing is *not* a plausible potential explanation of the chemical reaction, the presence of hydrogen *is*. Hence, (9a) is marginal to understanding the chemical reaction because it indicates a failure of the first mark of SEEing, consideration; (10a) does not bear this stigma. Similarly, changing our linguistic conventions would not alter the chemical reaction; removing lithium from the system would. As a result, we get more bang for our buck in including (10b) when comparing explanations than we do from including (9b). Hence, it appears that counterfactual reasoning is excessive if it plays no role in SEEing. But, of course, wherever it does play just such a role, it's redundant. Consequently, Grimm's counterfactual-reasoning ability should not depart from the EKS Model.

3.6.2. *Modal Apprehension*

Grimm's most radical departure from the received view not only denies that understanding is propositional knowledge, as he did with counterfactual-reasoning abilities, but replaces belief with a faculty that I call "modal apprehension."[23] He frequently treats modal apprehension and counterfactual reasoning as a package deal. I will argue that this is neither necessary nor desirable.

Grimm's account of modal apprehension borrows from Bonjour's (1998) account of *a priori* justification. As Grimm (2014, 334) interprets Bonjour:

> ... the metaphor of 'seeing' seems to involve ... something like an apprehension of how things stand in modal space. ... If this picture is correct, ... then for our purposes the important thing ... is that what is grasped or seen

---

[23] Instead of "modal apprehension," Grimm uses the word "grasping." As I've already earmarked the latter as a more generic term to indicate whatever cognitive relationship stands between an agent and the information that constitutes her understanding, I have baptized a more precise (and barbarous) label for Grimm's signature mental state.

... is not in the first instance a proposition but rather a modal relationship between properties (or objects, or entities) in the world.

This is precisely what I'll be calling "modal apprehension." On this view, understanding requires some kind of immediate acquaintance with modal relationships. If we take the visual metaphor seriously, then just as an apple impinges upon our visual faculties when seen, modal relationships impinge upon our modal faculties when apprehended. *A priori* knowledge involves apprehending modal relationships of the logical, mathematical, and conceptual variety, while understanding of empirical phenomena involves apprehending modal relationships of the causal and physical variety.[24] Furthermore, as the preceding quotation makes clear, modal relationships are not propositions. Hence, modal apprehension is not a kind of belief.

Let me put my cards on the table: like others (e.g., Devitt 2014), I find Bonjour's original proposal obscure, and Grimm's appropriation of it inherits the same worries. How do modal relationships "impinge" upon our modal faculties? Where is modal space located? If my prose sometimes betrays my incredulity, I hope I will be forgiven in advance. At any rate, this incredulity will serve as the last straw, not the first. My goal is to demystify Grimm's account of modal apprehension as much as possible. As I'll show, it still faces searching problems, even for those who lack my empiricist and naturalist leanings.

It sometimes appears that Grimm takes modal apprehension to be an explanation or description of counterfactual reasoning. The EKS Model provided a much simpler and straightforward explanation of counterfactual reasoning abilities: they fall out of scientists' knowledge of explanatory propositions, particularly in the context of SEEing. Open a science journal and you'll find various experiments, causal inferences, and statistical techniques that involve counterfactual reasoning of this sort, but you'll be hard pressed to find the intimate contact with modal reality that characterizes modal apprehension.

Consequently, modal apprehension's role in understanding must be independent of its connection to counterfactual reasoning since forging that connection appears altogether optional. However, so long as modal apprehension is separated from counterfactual reasoning abilities, it remains mysterious. More to the point, without the safety net of counterfactual reasoning, I fail to see how modal apprehension benefits our

---

[24] Sullivan (manuscript) offers a few different interpretations of Grimm's account of modal apprehension but also finds faults with his arguments for the non-propositional nature of understanding.

understanding. Why should we think that it improves understanding any more than Nate's eagle-eyed lip-reading in Section 3.4 described earlier?

How might we bring modal apprehension's features into sharper relief? Grimm's (2014, 334) clearest suggestion is that modal apprehension is a kind of *de re* knowledge. In other words, if Molly understands why the chemical reaction occurred via modal apprehension, then the following is true:

> (11) Molly knows *of* the chemical reaction's occurrence *how* different reactants would have changed it.

By contrast, Grimm claims that the kind of (testimonial/memorial) knowledge that affords little to no understanding is *de dicto*; e.g.:

> (12) Kate knows *that* the chemical reaction occurred because the oxygen was introduced.

However, it is hard to see how the *de re–de dicto* distinction is the right tool for Grimm's jobs. First, *de re* knowledge does not entail non-propositional knowledge. For instance, (11) boils down to the following:

> (13) For some *x*, *x* = the chemical reaction, and Molly knows *how* different reactants would have changed *x*.

If there is anything non-propositional about this, it is not because it is *de re*; it's because it's a kind of *know-how*. To see this, consider the following:

> (14) Molly knows *of* the chemical reaction's occurrence *that* different reactants would have changed it.

Quite clearly, this becomes:

> (15) For some *x*, *x* = the chemical reaction, and Molly knows *that* different reactants would have changed *x*.

Here, Molly's knowledge is propositional even though (15) is *de re*. To be sure, Grimm (2014, 335) anticipates this challenge:

> The basic idea here is . . . not that propositions have no role to play . . ., but rather that they play a secondary or derivative role . . . the primary object of *a priori* knowledge [and Grimm's account of understanding] is the modal reality itself that is grasped by the mind, and it is on the basis of this grasp that we then (typically) go on to assent to the proposition that describes or depicts these relationships.

However, in this passage, Grimm claims that we *base* our beliefs on modal apprehension. Consequently, modal apprehension is better regarded as a kind of *justification* rather than as some surrogate for belief. This would mean that Grimm's view is a far less radical departure from the received view – he would simply require understanding to be explanatory knowledge gained via modal apprehension.

However, even this more modest role for modal apprehension is a bridge too far. Assuming that modal apprehension is still a kind of *de re* knowledge, Grimm's claim is that this *de re* knowledge somehow justifies *de dicto* knowledge. However, the semantics of *de re* sentences speak against this. A sentence is *de re* just in case it allows for substitution of co-designating terms *salva veritate*; otherwise, it's *de dicto*. For instance, suppose that the following is true:

> (16) The chemical reaction = the oxidation of lithium.

Combined with (15), this entails the following:

> (17) Molly knows of the oxidation of lithium how different reactants would have changed it.

The move from (15) and (16) to (17) is possible precisely because Molly needn't possess the concepts *oxidation* or *lithium*. However, understanding of the chemical reaction improves if one can classify it as the oxidation of lithium. Consequently, if any knowledge or mental states justify understanding, they are likely to be *de dicto*.

Thus, the *de re–de dicto* distinction cannot do the work that Grimm asks of it. In other words, Grimm has failed to establish the following:

> UA3$_G$. If S_1 apprehends the modal relationship between p and q, S_2 does not, and they are otherwise identical, then, S_1 understands why p better than S_2.

Construed as *de re* knowledge, modal apprehension is thus excessive.

So perhaps, as suggested previously, modal apprehension has less to do with *de re* knowledge and is non-propositional primarily because it is a kind of *know-how*. For instance, Grimm (2014, 336) writes:

> In the case of knowledge of causes . . ., what [is] seen or grasped [is] how changes in the value of one of the terms of the causal relata would lead (or fail to lead) to a change in the other.

However, I fail to see why this could not be captured by our earlier treatment of counterfactual reasoning; e.g., by

> (18) Molly knows that the chemical reaction would have produced lithium hydride (rather than lithium oxide), had the reactant been hydrogen (rather than oxygen).

Although this is propositional (and, incidentally, *de dicto*) knowledge, Molly seems to know how changes in the reactant would lead to changes in the resulting compound. This would mean that modal apprehension construed as know-how is redundant. Indeed, it's hard to see how this proposal would be any less immune to the arguments raised against counterfactual reasoning in Section 3.6.1.

Thus, I conclude that modal apprehension is both obscure and excessive. I say this not only as a card-carrying empiricist and naturalist, but also as someone sympathetic to Grimm's broader engagement with the Classic Ability Argument. After all, he can still rebuff enablers by *either* developing his account of counterfactual reasoning abilities independently of his account of modal apprehension, as discussed in Section 3.6.1, or by expanding on a defense of the received view, similar to the one I advanced in Section 3.2.

3.7. Conclusion

I have argued that the EKS Model readily accommodates many of the abilities associated with understanding, and I have also used this chapter to highlight when other proposals have mischaracterized or misidentified the abilities that have figured in understanding. Pritchard's "why-how achievements," in which one gains a true belief about why something is the case based on one's grip of how the explanation goes, were largely seen to be excessive. Hills's "cognitive control" and Grimm's two accounts of grasping were more mixed, containing some excess and some redundancy.

I conclude this section by briefly noting that enablers are often motivated by dissatisfaction with the received view's treatment of the "grasping" that is supposed to be characteristic of understanding. If the preceding arguments are sound, that's a mistake. Grasping isn't special. In no context is it anything more than scientific knowledge of an explanation – and in many contexts, it's substantially less.

CHAPTER 4

Objectual Understanding

Often, exemplary understanding involves mastering a subject matter. For instance, the physicists from Chapter 2 didn't simply understand why the scattering results departed from the predictions of quantum electrodynamics and why the proton structure functions scaled. Additionally, they understood how the explanations of those phenomena fit within the larger context of particle physics. So, it's tempting to think that their understanding of particle physics – a subject matter – is far more precious than the piecemeal explanatory understanding to which I gave top billing.

Epistemologists have designated this mastery of a subject matter with the barbarous label of "objectual understanding." If objectual understanding is *sui generis*, then an interesting kind of understanding that is not simply explanatory knowledge is worthy of philosophical attention. In principle, this need not run afoul of the received view, which, as you'll recall, only asserts that:

> *S* understands why *p* if and only if there exists some *q* such that *S* knows that *q explains why p.*

Since we needn't characterize subject matters (such as particle physics) in terms of why-questions, one could consistently hold that the received view is the proper account of understanding-why while also insisting on a distinctive theory of objectual understanding.

That being said, let's not forget the stance that animates the received view: that an epistemology of scientific explanation exhausts what is philosophically notable about understanding. Such a stance courts suspicion about the baptism of new kinds of understanding that purport to outstrip explanatory knowledge. This brings us to the Classic Objectual Question, first mentioned in Chapter 1, which asks:

- Is objectual understanding the same as explanatory understanding?

In principle, one could answer this question negatively while also adopting an irenic view in which objectual and explanatory understanding peacefully coexist. However, I aim to disrupt this happy scene: aside from terminological convenience, anything noteworthy about objectual understanding can be replicated by explanatory understanding without loss. Conversely, all accounts of objectual understanding face liabilities that explanatory understanding avoids. Hence, on the whole, we would be well served to rid ourselves of heavyweight notions of objectual understanding.

However, to efface the differences between explanatory and objectual understanding, I will not be characterizing the former in terms of the received view. Rather, I will use its successor, the Explanation-Knowledge-Science (EKS) Model, which I have developed in earlier chapters:

(EKS1) S_I understands why p better than S_2 if and only if:
    (A) *Ceteris paribus*, S_I grasps p's explanatory nexus more completely than S_2; or
    (B) *Ceteris paribus*, S_I's grasp of p's explanatory nexus bears greater resemblance to scientific knowledge than S_2's.

(EKS2) S has minimal understanding of why p if and only if, for some q, S believes that *q explains why p*, and *q explains why p* is approximately true.

As a result of turning from the received view to the EKS Model, two "objectualist" foils emerge. First, "robust objectualists" claim that objectual understanding requires *more* than explanatory understanding (Carter and Gordon 2014; Elgin 2007; Kvanvig 2003). They claim that explanations cannot provide the breadth and systematicity typically associated with objectual understanding. Because the EKS Model countenances degrees of understanding, the Classic Objectual Question isn't the best way to assess robust objectualism: as I'll argue later, it is trivial that *modest* understanding-why requires less than *demanding* objectual understanding. Rather, we should be asking what I called the Updated Objectual Question:

- Are comparably demanding cases of objectual and explanatory understanding identical in all philosophically important ways?

I shall argue that robust objectualists have failed to provide compelling reasons to answer this question negatively.

The other objectualist camp – whom I shall call "austere objectualists" – take objectual understanding to require *less* than explanatory understanding (Gijsbers 2013, 2014; Kelp 2015; Kvanvig 2009b). They point to the non-explanatory understanding we get from classifications, discoveries, and non-deterministic systems that resist explanation. They, too, are best seen as answering the Updated Objectual Question negatively, though their foil will be the kind of *minimal* explanatory understanding described by EKS2. I will argue that a slight modification to the EKS Model readily douses any motivations that would spark austere objectualism.

I begin by setting the stage (Section 4.1). I then canvass the four leading ways in which objectual understanding is alleged to differ from explanatory understanding, arguing that we can replicate these features without harm using only the resources featured in my account of explanatory understanding (Sections 4.2 to 4.5).

4.1. The Lay of the Land

Even if we restrict ourselves to empirical phenomena, there is no shortage of kinds of understanding. Consider the many ways that we can understand an imploding building: we can understand the implosion, demolitions, *that* the building imploded, *which* building imploded, *when* the implosion occurred, *where* the implosion occurred, *what* happened when we pushed the shiny red button, and, of course, *how* and *why* it imploded. The first two items on this list are instances of *objectual* understanding; the third, *propositional* understanding; and the rest, *interrogative* understanding. Among the kinds of interrogative understanding, only the last two are instances of *explanatory* understanding.

It's fair to say that propositional and non-explanatory interrogative understanding have not sparked challenges to the received view. I suspect that this is because of a widespread assumption that propositional understanding and interrogative understanding are quite similar, if not identical, to their respective kinds of knowledge. This suggests that understanding-why is a kind of knowledge-why, which is precisely what the received view proposes.

Not surprisingly, then, many critics of the received view focus on the relationship between objectual and explanatory understanding. But what exactly *is* objectual understanding? Its paradigmatic expressions involve "understands" followed by a noun phrase (e.g., "Michelle understands the New York Subway system" or "Tom understands ecology"). Hereafter, typical cases of objectual understanding have the form "*S* understands *O*," where "*S*" names a person, and "*O*" is a noun phrase referring to a subject matter, body of information, or domain.[1] Thus, the New York Subway system might be seen as a subject matter, body of information, or a domain; ditto for ecology. So, objectualists deny that the understanding of a domain is identical to knowledge of an explanation. Pre-theoretically, it certainly seems plausible to think that, e.g., knowing why an individual species went extinct is not the same as understanding the field of ecology.

This raises the Classic Objectual Question:

- Is objectual understanding the same as explanatory understanding?

Let *classic objectualism* be any position that answers the Classic Objectual Question negatively. We can then subdivide objectualism as we suggested earlier:

- *Classic Robust Objectualism* (CRO): Some instances of explanatory understanding are not instances of objectual understanding.
- *Classic Austere Objectualism* (CAO): Some instances of objectual understanding are not instances of explanatory understanding.[2]

Note that these doctrines are not mutually exclusive; an objectualist could endorse both.

Many objectualists have endeavored to answer the Classic Objectual Question. However, so formulated, it obscures the important role of *degrees* of understanding in assessing objectualist arguments. For example, it would clearly be a cheap victory to show that someone who has minimal understanding of a very narrowly circumscribed phenomenon differs substantially from someone who has top-shelf understanding of an expansive subject matter. If we really need a standalone concept of objectual

---

[1] I will use the terms "subject matter," "body of information," and "domain" interchangeably.

[2] More precisely:

- CRO states that $\exists S \exists p(($"*S* understands why *p*" is true$)$ & $\forall O(($*p* is about *O*$) \rightarrow ($"*S* understands *O*" is false.$)))$
- CAO states that $\exists S \exists O(($"*S* understands *O*" is true$)$ & $\forall p(($*p* is about *O*$) \rightarrow ($"*S* understands why *p*" is false.$)))$

Here, I have added that the explanandum *p* must be about the subject matter *O* for something more lifelike. Otherwise, robust objectualists would be committed to the existence of explanatory understanders who had no understanding of *any* subject matter whatsoever, even subject matters that have nothing to do with *p*; austere objectualists would face a mirror image situation. For simplicity of exposition, this addendum will figure only implicitly in my discussion throughout the chapter.

understanding, then the *Updated* Objectual Question ought to be answered in the negative:

- Are comparably demanding cases of objectual and explanatory understanding identical in all philosophically important ways?

The Updated Objectual Question requires two points of clarification. First, when are objectual and explanatory understanding "comparably demanding"? This depends mightily on whether we are assessing robust or austere objectualism. The "updated robust objectualist" must claim that some instances of objectual understanding outstrip *any* degree of explanatory understanding.[3] By contrast, the "updated austere objectualist" must claim that some instances of objectual understanding fall short of even the most modest instance of explanatory understanding (i.e., *minimal* understanding, as defined by EKS2).

The Updated Objectual Question requires a second clarification: what makes for a "philosophically important" difference or similarity? Objectualists have made four claims that fill this gap:

- *Robust Breadth*: Objectual understanding is of a comprehensive body of information; explanatory understanding is narrower or more piecemeal (Carter and Gordon 2014; Elgin 2007; Kelp 2015; Kvanvig 2003). Hence, some instances of objectual understanding involve information or answers to questions that no degree of explanatory understanding could replicate (Carter and Gordon 2014).
- *Austere Breadth*: Objectual understanding is of a comprehensive body of information, including non-explanatory information. Hence, even in the absence of minimal explanatory understanding, objectual understanding can be achieved by grasping non-explanatory information or answering a non–explanation-seeking question (Kelp 2015).
- *Robust Coherence*: Some instances of objectual understanding involve coherence-making relationships that no degree of explanatory understanding could replicate (Carter and Gordon 2014; Elgin 2007).

---

[3] This includes the *most* demanding instances of explanatory understanding, what I called "ideal understanding" in earlier chapters:

> *S* ideally understands why *p* if and only if it is impossible for anyone to understand why *p* better than *S*.

As discussed in Chapter 1, this is parasitic upon EKS1's account of better understanding. Having said this, the objections to robust objectualism developed later have implications for ideal understanding, but do not explicitly appeal to it.

- *Austere Coherence*: Even in the absence of minimal explanatory understanding, objectual understanding can be achieved by grasping coherence-making relationships (Kvanvig 2009b).

These four claims give "updated objectualism" a voice. I also assume that they exhaustively describe any *philosophically important* differences that could distinguish objectual from explanatory understanding.[4] Hence, if any one of them were borne out, then objectualism would be vindicated: there would be philosophically important differences between objectual understanding and explanatory understanding.

4.1.1. Quasi-Explanationism

Note that one could embrace any of these four differences while still accepting a separate account of understanding-*why*. However, the larger aim of this book is to undercut the motivations for thinking that anything about understanding exceeds the epistemology of scientific explanation.[5] To that end, I will argue that none of these four claims pans out. Aside from its grammar and some empty slogans about "subject matters," objectual understanding bears no philosophically important differences from explanatory understanding. Thus, an affirmative answer to the Updated Objectual Question asserts the following:

> *Explanationism*: Nothing of philosophical importance is lost if all instances of objectual understanding are treated as instances of explanatory understanding.

Strictly speaking, I do not endorse explanationism, though I endorse its near neighbor, viz:

> *Quasi-Explanationism*: Nothing of philosophical importance is lost if all instances of objectual understanding are treated as **(one or more)** instances of:
> (a) explanatory understanding, or
> (b) **being on the right track** to having explanatory understanding.

Here, to treat an instance of objectual understanding as *one* instance of explanatory understanding is to claim, for instance, that "Tom understands ecology" is simply shorthand for, e.g., "Tom understands how

---

[4] Obviously, if other objectualists should propose new candidates in addition to these four claims, further discussion would be needed.
[5] See Chapter 1 for details.

organisms and environments are interrelated."[6] Since the complement of
the latter expression is an embedded explanation-seeking question, it
counts as an instance of explanatory understanding. Both explanationism
and quasi-explanationism claim that these two sentences pick out cognitive
achievements that are identical in terms of their breadth and coherence.
Hence, if either of these positions is defensible, then we gain nothing by
introducing a new concept of objectual understanding.

While explanationism requires a one-to-one mapping between instances
of objectual and explanatory understanding, my preferred doctrine, quasi-
explanationism, also permits one-to-many mappings. To treat an instance
of objectual understanding as *multiple* instances of explanatory understanding is to claim, for instance, "Tom understands ecology" is simply
shorthand for, e.g., "For many p about ecology, Tom understands how/
why p." This feature of quasi-explanationism is particularly effective in
rebutting robust objectualism, for frequently nothing is lost by thinking of
"robust" objectual understanding as a stockpile of explanatory understanding. Unlike explanationism, this quasi-explanationist maneuver may be
available even when no single instance of explanatory understanding
suffices to replicate understanding of a subject matter.

Quasi-explanationism also departs from strict explanationism in allowing some instances of objectual understanding to be treated as one or more
instances of "being on the right track" to having explanatory understanding. In this case, to claim, for instance, that "Tom understands ecology" is
simply shorthand for, e.g., "For some p about ecology, Tom is on the right
track to understanding how/why p." Here, "being on the right track"
means that one does not know the proper answer to a relevant explanation-seeking question, but one has information that would be useful in
acquiring such knowledge. This proves to be especially effective in defusing
austere objectualism, since being on the right track does not require that
one has already arrived at the desired destination.

To match objectualists in terms of barbarous labels, let's say that someone who is on the right track to having explanatory understanding possesses "(explanatory) *proto-understanding*," and define it thus:

> (PU) S has proto-understanding of why p if and only if:
> (1) S does not understand why p, and

---

[6] I do not claim any tidy translation manual for this, largely because I am skeptical that objectual
understanding comes anywhere close to picking out a natural kind. In this case, the suggested
recasting of objectual in terms of explanatory understanding was inspired by looking at a few
different definitions of ecology.

(2)   For some proposition q,
    a. q plays explanatory role R with respect to p; i.e., Rpq, and
    b. S's mental state toward Rpq is the same kind of mental state toward Rpq that figures in scientific knowledge of why p.

While *explanations* are the prime movers in understanding-why, *explanatory roles* drive proto-understanding. To get a better sense of what these roles are, recall that on my own EKS Model, explanatory understanding improves in proportion to the explanatory information grasped and the resemblance of that grasp to scientific knowledge, viz:

(EKS1) S_1 understands why p better than S_2 if and only if:
(A)   *Ceteris paribus,* S_1 grasps p's explanatory nexus more completely than S_2; or
(B)   *Ceteris paribus,* S_1's grasp of p's explanatory nexus bears greater resemblance to scientific knowledge than S_2's.

As I have throughout, I will call (A) the "Nexus Principle" and (B) the "Scientific Knowledge Principle." Roughly, explanatory roles are propositions that play a part in improving understanding in accordance with these two principles.

More precisely, a proposition plays a *direct* explanatory role with respect to p if and only if it is (or describes) one of the following components in a correct explanation of p: explanans, explanandum (i.e., p), explanatory relation, or an explanatory presupposition. It plays an *indirect* explanatory role if it plays no direct role but figures in scientific knowledge of a correct explanation of p. With this, we can see why proto-understanding amounts to being on the right track: when inquirers grasp only a few explanatory roles, they have bits and pieces of explanatory understanding but not enough to actually provide a correct explanation.

4.1.2.   *Death in the Bahamas*

Let's get a better feel for what these roles involve. Trivially, propositions describing explanations will involve all of our direct explanatory roles. For instance, consider the following example from ecologists Schoener, Spiller, and Losos (2001):

(1)   Predation by northern curly-tailed lizards and the occurrence of Hurricane Floyd caused some Bahamian islands' brown anole populations to go extinct in 1999.

This correct explanation recruits the explanans (concerning curly-tailed lizards and Hurricane Floyd); the explanandum, *some Bahamian islands' brown anole populations went extinct in 1999*; a description of the causal mechanisms and structures that led to the extinctions (e.g., predation), and various presuppositions (see my discussion of propositions (5) and (6), for an example).

However, proto-understanding occurs only in the absence of explanatory understanding of why *p*. This immediately points us toward the direct explanatory roles that will be most relevant for our discussion, e.g.:

(2) Some Bahamian islands' brown anole populations went extinct in 1999.
(3) Hurricane Floyd hit the Bahamas in 1999.
(4) Northern curly-tailed lizards were present on some Bahamian islands in 1999.
(5) Northern curly-tail lizards prey upon brown anoles.

However, simply having true beliefs about (2)–(5) does not, according to PU, provide proto-understanding of (say) why some Bahamian islands' brown anole populations went extinct in 1999. Additionally, one must recognize the explanatory roles that these statements play. So, for instance, while proto-understanding can't be attained simply by correctly believing that some Bahamian anole populations died off in 1999, i.e. (2), such a belief is a springboard for proto-understanding if a person has a true belief that:

(6) Something caused some Bahamian islands' brown anole populations to go extinct in 1999.

In this case, the inquirer recognizes that the extinction is an explanandum and thereby has proto-understanding of the anole extinction. Parallel points apply to explanantia, e.g. (4), wherein proto-understanding might consist of the true belief that:

(7) The presence of Northern curly-tailed lizards on some Bahamian islands in 1999 caused something to happen to the brown anoles on those islands.

Additionally, explanations frequently have interesting presuppositions (Garfinkel 1981; Hitchcock 1999; Risjord 2000; Sober 1986). We might view (5) as just such a presupposition. Perhaps its corresponding "role-statement" would be:

(8) If northern curly-tailed lizards affect the population of brown anoles, then it is because northern curly-tail lizards prey upon brown anoles.

As with (6) and (7), a true belief in (8) furnishes proto-understanding.

In addition to these direct explanatory roles, several propositions play *indirect* explanatory roles. These are propositions that are not part of an explanation but are part of scientific knowledge of an explanation. On this front, recall that I departed from the received view's equation of understanding-why with *mere* knowledge of an explanation, opting instead to cast understanding-why in terms of *scientific* knowledge of an explanation. I then defined this latter kind of knowledge as follows:

> *S* has scientific knowledge that *q explains why p* if and only if the safety of *S*'s belief that *q explains why p* is because of her scientific explanatory evaluation.

As discussed at greater length in the previous chapters, I take scientific explanatory evaluation (SEEing) to consist of three features: consideration of plausible potential explanations of a given phenomenon, comparison of those potential explanations using the best available scientific evidence and methods, and formation of attitudes on the basis of those comparisons.

By far the most important indirect explanatory role is when a proposition is a prediction of an explanation (whether potential or actual).[7] Paradigmatically, such propositions are answers to what-if (-things-had-been-different) questions, as described by Woodward (2003). For instance, Schoener et al.'s research allows us to answer questions such as, "What would happen to a Bahamian island's brown anole population if there were no northern curly-tailed lizards on it just before a hurricane struck?" and "What would happen to a Bahamian island's brown anole population if a hurricane wiped out all of the northern curly-tailed lizards?" (In both cases, the answer is: "The brown anole population would have survived.") Furthermore, as we saw in the previous chapter, answers to what-if questions enhance our understanding when they play a clear role in SEEing; when they don't, their contribution to our understanding is far less clear (see Section 3.6.1).

Another important indirect explanatory role is played by propositions describing the evidence that scientists have for their explanatory model. Such evidence figures in the second stage of SEEing (comparison). For instance, here is one piece of evidence:

---

[7] Depending on one's view of contrastive explanation, answers to what-if questions might also be part of the explanans and explanandum, or they might be presuppositions. In this case, they play direct explanatory roles but are reined in by the explanandum of interest rather than by SEEing. Given the arguments of the previous chapter, I prefer to regard them as playing indirect explanatory roles.

(9) Bahamian island A13 had no northern curly-tails before or after
Hurricane Floyd and had a similarly sized anole population in
October 2000 as it did in April 1998.[8]

As with direct explanatory roles, it is not enough simply to believe this.
One must also identify its role. In this case, that might require some further
cognizance of the larger design of the study (e.g., when Hurricane Floyd
occurred) and that studying other islands with northern curly-tailed lizards
is needed to ascertain the effects of northern curly-tails on anole
populations.

Another indirect explanatory role is the relationship between two or
more correct explanations of the phenomenon of interest. Indeed, one of
Schoener et al.'s most interesting discoveries is the *interaction* between the
two explanations that they considered: how the existence of predators (the
northern curly-tailed lizard, *Leiocephalus carinatus*) and the effects of a
catastrophic event (Hurricane Floyd) were both necessary for the extinction of the prey population (the brown anole, *Anolis sagrei*) to occur.
Specifically, Schoener et al. introduced *L. carinatus* to six of the twelve
islands that they studied, all of which contained *A. sagrei*. Prior to the
hurricane, *A. sagrei* populations on the experimental islands (i.e., those
with *L. carinatus*) decreased by half. After the hurricane, the only islands to
see their *A. sagrei* populations go extinct were those that still had *L.
carinatus* populations after the hurricane. (*A. sagrei* populations rebounded
on the experimental islands that witnessed *L. carinatus* extinctions as a
result of the hurricane.)

Finally, note that proto-understanding is not limited to *belief*. This is
particularly apparent when it comes to indirect explanatory roles. Indirect
explanatory roles should play the same role in a proto-understander's
mental life as they would in scientists' best feats of explanatory evaluation
(while holding the explanandum fixed). For instance, in some cases, it is
enough to have *considered* a plausible potential explanation and perhaps to
have *rejected* it on the basis of sound comparison. Similarly, methodological principles may not be objects of belief but may simply be *assumed*.

These non-doxastic attitudes can provide proto-understanding in at
least three different ways. First, simply considering plausible potential
explanations of a phenomenon enhances one's objectual understanding.
The scientists who studied the anole population did just this. While the
scientists grasped the explanation of why the anoles went extinct on certain

---

[8] The scientists coded the twelve islands they studied using letters and numbers: 1, A13, A17, N15, X3,
Z5, A18, N1, X10, Z3, Z4, Z1.

islands (see (1) above), they did not know how the presence of *L. carinatus* interacted with Hurricane Floyd's occurrence to wipe out anole populations on certain islands. Instead, they only considered three potential hypotheses about this interaction. Interestingly, all three hinge on a crucial observation: no mature members of either population survived the hurricane and its resulting 3-meter storm surge. First, immediately after the hurricane, *L. carinatus* hatchlings may have preyed on *A. sagrei* hatchlings. Second, prior to the hurricane, *A. sagrei* retreated to higher branches, where *L. carinatus* were less inclined to climb. However, laying eggs in these higher places is less secure, especially during a hurricane. Third, the rate of egg production per female anole may have been reduced by the presence of northern curly-tailed lizards. As they admit, there is insufficient evidence to believe any one of these explanations. Nevertheless, by considering these plausible potential explanations, they are on the right track to understanding how predation and the hurricane could interact.

A second non-doxastic way that indirect explanatory roles contribute to objectual understanding is by being able to disconfirm otherwise-plausible potential explanations through sound scientific method and evidence. For instance, suppose that Yvette and Zachary discover that the poor brown anole is on the wane on a certain Bahamian island. Zachary does not consider why this is so. Yvette considers that it might be a food shortage or an epidemic but, upon further inspection of the anoles' habitat, can safely rule these out. However, neither of our inquirers ever comes to know of the northern curly-tails' voracious appetite for brown anoles or of Hurricane Floyd. *Ceteris paribus*, Yvette is still closer than Zachary to understanding the decrease in the Bahamian population of the brown anole, just as PU suggests.[9]

Third, propositions about the methodological prescriptions used in these comparisons play indirect explanatory roles and intuitively grasping them puts us on the right track to understanding why. For instance, suppose that neither Victor nor Wendy entertains beliefs as to why the brown anole population took a hit as a result of hungry curly-tailed lizards and Hurricane Floyd. Both encounter the same pieces of (inconclusive) evidence for this explanatory hypothesis, but only Wendy has enough know-how with respect to causal inference to recognize that the evidence she's encountered is inconclusive. Wendy is closer to understanding why the brown anole populations went extinct than Victor.

[9] Strictly speaking, PU does not countenance degrees of proto-understanding. A slight variant of EKS1 would readily fill in this lacuna. I leave this as an exercise for the reader.

In sum, objectualists claim that breadth and coherence distinguish our understanding of subject matters from our explanatory understanding. Quasi-explanationism claims that these differences are illusory: objectual understanding boils down to having lots of explanatory understanding, garden-variety explanatory understanding, or being on the right track to having explanatory understanding. In the balance of this chapter, I show how quasi-explanationism handles these various objectualist proposals. By the end of this romp, I will have gotten all of the objectualists' mileage with only explanation and scientific knowledge in my tank.

4.2.   Robust Breadth

One obvious difference between subject matters and explanations is that we tend to think of the former as more comprehensive than the latter. Intuitively, mastering a larger body of information is a greater cognitive achievement, and such achievements are philosophically important. Hence, we might think that objectual understanding's greater *breadth* justifies objectualism. Carter and Gordon (2014, 8) suggest a useful way of thinking about this:

> ... those in possession of rich understanding of a subject matter are able to answer a broader class of relevant questions with ease.

This has intuitive pull. For instance, if Michelle understands the New York Subway system, she can answer a wide variety of questions about the subway system. Carter and Gordon only argue for classic robust objectualism. As mentioned earlier, framing the issue in this way is not profitable. To that end, I begin by highlighting the shortcomings of this approach (Section 4.2.1). I then argue that, although using breadth to answer the Updated Objectual Question is an advance, it still offers no refuge from quasi-explanationism (Section 4.2.2).

4.2.1.   The Classic Breadth Argument

A person might have explanatory knowledge but be unable to answer many questions with that knowledge. Carter and Gordon (2014) offer just such an argument. For instance, they present the following thought experiment:

Forrest and Emily
Forrest is a novice fire officer and not very bright. Nevertheless, in doing his job, he correctly explains that a local house burned down because of faulty wiring, but can piece together little else about the events surrounding the fire.

By contrast, Emily is a leading expert in exothermic chemical reactions, has had a team study the house's wiring, examined photographs of the frequency spectrum of the flames, and conducted a detailed stoichiometric analysis of the fire site, and as a result, can provide a much more detailed, accurate, and comprehensive explanation of why the same house burned down.[10]

Carter and Gordon conclude that although both Forrest and Emily understand why the house burned down, Emily's understanding is superior because only she has broad, "rich" objectual understanding.

Admittedly, Carter and Gordon's concern is not, in the first instance, about the Classic Objectual Question. Rather, they argue that the *value* of objectual understanding is more fundamental than the *value* of explanatory understanding:

> Cases we might be inclined to call "rich understanding-why" are, after all, cases we are able to distinguish as such *only because* of the valuable properties of the objectual understanding that will already be present. In this respect, the value of the richness of the latter has a kind of explanatory priority. (Carter and Gordon 2014, 8; emphasis added)

As this quotation makes clear, Carter and Gordon believe that there are valuable properties that objectual understanding has by default but that understanding-why only has contingently. However, if there are *valuable* properties that enjoy this status, then there are *some* properties that understanding-why only has contingently but that objectual understanding enjoys by necessity. Hence, all told, Carter and Gordon are committed to robust objectualism, even if it's not their chief concern.

Having said this, Carter and Gordon's argument is flawed. It appears to run as follows:

Classic Breadth Argument

CB1. Emily's objectual understanding requires her to answer a broad class of questions about a subject matter.

CB2. Forrest's explanatory understanding does not require him to answer a broad class of questions about a subject matter.

CRO. ∴ *Classic Robust Objectualism:* Some instances of explanatory understanding are not instances of objectual understanding. (CB1, CB2)

However, this argument is not valid. In particular, Forrest's low-grade explanatory understanding of why the house burned down may fail to replicate Emily's high-grade objectual understanding of exothermic

---

[10] This is a paraphrase.

reactions[11] while still replicating *some other* instance of objectual understanding: namely, low-grade objectual understanding of the house's burning down. If this is so, then the example fails to support robust objectualism. So, to get something approaching a valid argument, Carter and Gordon's example requires two revisions.

First, to show that Emily's understanding is of greater breadth in virtue of being objectual, we need "fair comparisons" between explananda and maximally similar subject matters.[12] By contrast, Carter and Gordon argue for robust objectualism by pointing out that Forrest (explanatorily) understands why the house burned down even though he does not (objectually) understand exothermic reactions. Quite clearly, even the best understanding of why a house burned down is not going to be the stiffest competition to understanding exothermic reactions since house fires and exothermic reactions are not all that similar. Consequently, we should be reluctant to draw any claims about robust objectual understanding's distinctness since it's no surprise that explanatory understanding of *some* phenomenon can be decoupled from objectual understanding of *some very different* subject matter. This is a far cry from showing that explanatory understanding of *any* phenomenon cannot capture objectual understanding of some subject matter, which is what classic robust objectualism requires. This suggests that we compare Forrest's *explanatory* understanding of *why the house burned down* with Emily's *objectual* understanding of *the house's burning down.*

However, even this is not a fair comparison. A person can answer more or fewer questions about a subject matter, so breadth clearly admits of degrees. After all, Forrest can quite clearly answer the question, "Why did the house burn down?" As such, if broad objectual understanding were treated as an all-or-none affair, then we are owed some principled cut-off point between answering *many* questions about a domain and merely answering *some* of these questions. Carter and Gordon have offered no such dividing line, so the distinction between low-grade explanatory understanding and low-grade objectual understanding appears arbitrary. So, our second revision requires that we compare low-grade explanatory understanding to low-grade objectual understanding.

---

[11] Carter and Gordon also mention other subject matters, e.g., stoichiometry. For ease of exposition, I focus only on exothermic reactions. Adding to the list of domains that Emily objectually understands would not affect my arguments.

[12] I've learned much about fair comparisons from Brogaard (manuscript), though she's more interested in fair comparisons between different kinds of knowledge and understanding, while I'm more interested in fair comparisons between explanatory and objectual understanding.

Combined, these two considerations suggest that there is no principled reason to deny that Forrest has objectual understanding of the house's burning down. However, this appears to be entailed by (if not equivalent to) his explanatory understanding of why the house burned down. Indeed, the following sentence borders on the incoherent:

> Forrest understands why the house burned down, but he doesn't understand the house's burning down.

Thus, the example cannot fund an argument for robust objectualism: for all that Carter and Gordon have argued, Forrest's objectual understanding requires answering the same range of questions as his explanatory understanding. This, of course, is in the spirit of quasi-explanationism.

In sum, so long as understanding is treated as an all-or-none affair, the Classic Breadth Argument relies on illicit comparisons between explanatory and objectual understanding. Indeed, it appears that as Forrest's *explanatory* understanding improves, so does the number of questions that he can answer. So, for any degree of objectual understanding, there is prima facie reason to think that some comparable instance of explanatory understanding is just as broad. However, this is merely prima facie. To get to the heart of the matter, we need to put degrees of explanatory and objectual understanding under the microscope – we need to address the Updated Objectual Question.

4.2.2. *The Updated Breadth Argument*

Confession: I have racked my brain trying to come up with a workable notion of "comparable" cases of explanatory and objectual understanding. After much intellectual anguish, the following strikes me as a fair test: in cases where we are evaluating robust objectualism, we begin with two agents who have identical explanatory understanding. Since robust objectualism is consistent with the idea that explanatory understanding is necessary but not sufficient for objectual understanding, this is fair. We then see if there are any ways of improving objectual understanding that do not conform with EKS1. In other words, are there any ways of improving objectual understanding that don't boil down to grasping more explanatory information in ways that resemble scientific knowledge? Here's a simple example of the kinds of case I have in mind:

> Initially, Gabe and Holly have the exact same understanding of why the house burned down. Later, only Holly learns the answer to one further question about the house fire, while Gabe's understanding remains

unchanged. Furthermore, Holly's newly acquired information plays no role in explaining why the house burned down.

The robust objectualist holds that, in such cases, while both Gabe and Holly understand why the house burned down to the same degree, Holly (objectually) understands the house's burning down in a way that no degree of explanatory understanding could replicate. More precisely:

Robust Breadth Argument

RB1.   Objectual understanding can be improved upon by answering more questions about a domain.

RB2.   Some answers to questions about a domain play no explanatory role.[13]

RB3.   The only way to improve explanatory understanding is by playing an explanatory role.

RB4.   ∴ *Robust Breadth:* Some instances of objectual understanding involve answers to questions that no degree of explanatory understanding could replicate. (RB1–RB3)

Hence, as a committed quasi-explanationist, my job is to tell you where this argument falls short. Given that RB3 is a consequence of the EKS Model, I readily accept it. So, as I see it, the Robust Breadth Argument's other premises are its soft spots. More precisely, I will argue that the examples that make RB2 true leave RB1 unjustified. Hence, breadth is not the proper channel for defending updated robust objectualism.

In effect, robust objectualists face a dilemma similar to the one inflicted upon the "enablers" from the preceding chapter. Either answering a question about a subject matter plays an explanatory role or it does not. If it does play such a role, Robust Breadth (RB4) is *redundant*: we could just as well get it using the EKS Model. Moreover, any answers of this sort will flout RB2, so the improvement in question provides no basis for saying that objectual and explanatory understanding improve in different ways. If, on the other hand, the improvement in question plays no explanatory role, then I will argue that RB1 is highly dubious. All told, the dilemma raises the quasi-explanationist suspicion that objectual understanding does not stand freely of explanatory understanding. To that end, I shall now argue that breadth succumbs to the dilemma of redundancy and excess.

---

[13] While the term "explanatory role" was used earlier to describe a foil to austere objectualism, note that it works just as well in this context. After all, according to the Nexus Principle, direct explanatory roles are propositions that, when grasped, enhance our understanding; ditto for the Scientific Knowledge Principle and indirect explanatory roles.

Begin with redundancy. Carter and Gordon's paradigm case of the kinds of "relevant questions" characteristic of breadth are what-if questions. For instance, Holly might be capable of answering the question, "What would have happened if the house's shingles were made of fiberglass instead of rubber?" In the context of the Robust Breadth Argument, this is curious because answers to what-if questions are the signature kind of *explanatory* information. More precisely, I've argued that these sorts of questions play indirect explanatory roles; i.e., they enable us to have scientific knowledge of our explanations.

More generally, breadth fails to be a distinguishing feature of robust objectual understanding so long as the "relevant questions" that can be answered are demands for explanatorily relevant information. A person who can answer these questions conforms to the Nexus Principle. Hence, insofar as only answers to what-if questions are characteristic of breadth, explanatory understanding will be as broad as objectual understanding (i.e., RB2 is false). Consequently, robust objectualists will have to provide more nuanced arguments when invoking breadth.

Nor are what-if questions the only kinds of questions for which charges of redundancy will arise. For instance, suppose that the breadth of Holly's richer objectual understanding is captured by her ability to answer questions such as:

- How did the faulty wiring cause the fire?
- Which part of the wiring was faulty?

Answering how the faulty wiring caused the fire and which part of the wiring was faulty provides a more detailed mechanism of the fire's occurrence. Since mechanistic details are, *ceteris paribus*, explanatorily relevant (Craver 2007; Machamer, Darden, and Craver 2000), this enhances our explanatory understanding of why the fire occurred. All of this, of course, is congenial to quasi-explanationism.

Additionally, some questions will have answers that play other kinds of indirect explanatory roles. For instance, consider the following questions:

- What were the burn patterns near the breaker box like?
- What was the frequency spectrum of the flames?
- What did the stoichiometric analysis of the fire site reveal?

Answers to these questions won't be explanations of why the house burned down, but they describe pieces of scientific evidence used to evaluate such explanations. Hence, the more of these questions that one can answer, the more one resembles a scientist. Thus, there are many ways to deny RB2: gains in breadth frequently line up with grasping more explanatory roles.

Of course, if the robust objectualist can find answers to questions that play *no* explanatory role, then the game is not yet up, and she will have avoided charges of redundancy. I now want to throw some additional heft on to that burden of proof because this leads to the robust objectualist's second problem – the problem of excess. Without the guidance of explanatory considerations, robust objectualists have no way to discriminate between answers to questions that bolster our objectual understanding and those that amount to little more than "epistemic hoarding" (i.e., the haphazard gathering of trivia about a topic).

For instance, some questions about the house's burning down are explanatorily irrelevant; e.g.:

- How does one say, "The house burned down" in Mandarin?
- What was the middle name of the firefighter who extinguished the fire?

However, even if someone could answer these questions about the house burning down, I assume that one's *objectual* understanding of the house fire is not thereby enhanced. That is, even robust objectualists would deny the following in such cases:

> GH1. Gabe and Holly have identical understanding of why the house burned down, but Holly has greater objectual understanding of the house's burning down.

If the only difference between Holly and Gabe needed to justify GH1 is, e.g., the former's knowledge of firefighters' middle names, then robust objectualism licenses precisely the kind of epistemic hoarding (i.e., the indiscriminate amassing of tangentially related information) that smacks of excess. So, absent further qualifications, RB1 is highly dubious.

Since it will prove instructive later, let us consider a more plausible way of interpreting this example:

> GH2. Gabe and Holly have identical understanding of why the house burned down, but only Holly knows that the middle name of the firefighter who extinguished the fire is "Josephine."

If this is all there is to breadth, then there's no sense in which we need objectual understanding. What is understood is understood only explanatorily, and the rest can be covered by other shopworn cognitive statuses such as knowledge.[14]

But suppose we choose a less goofy question. For instance, suppose that Holly can answer the question, "When did the fire occur?" In some cases, knowing when the fire occurred will play an explanatory role (e.g., knowing that the fire occurred during an especially dry season). However, in this case, the understanding provided is redundant, for surely the dryness of the season helps to explain *why* the fire occurred. What about cases where knowing when the fire occurred plays no obvious explanatory role (e.g., if Holly knew that the fire occurred on June 11, 2014)? The robust objectualist might cite such an example as evidence for GH1.

But it seems to me that, if this were the *sole* difference between Gabe and Holly, then the situation admits of parallel treatment as GH2 above. The only difference is that Holly's hoarding is less perverse; i.e., we could just as well say that:

> GH3. Gabe and Holly have identical understanding of why the house burned down, but only Holly knows when the fire occurred.

In other words, there's no obvious loss of meaning or philosophical insight in describing Holly's greater understanding in terms of GH1 rather than GH3. As such, this example cannot support robust objectualism any better than Holly's knowledge of the firefighter's middle name. Hence, the goofiness of the question doesn't really bear on the issue of excess: robust objectualists still haven't shown us why we need a separate concept of objectual understanding. Indeed, so long as breadth is the only difference between objectual and explanatory understanding, the preceding treatment can be repeated *ad nauseum*.

Furthermore, there are good reasons to think that, even in this case, knowing when the fire occurred is compatible with my quasi-explanationist approach. In the current context, the key quasi-explanationist claim is that one *or more* instances of explanatory understanding amount to an instance of objectual understanding. For example, if only Holly can explain, e.g., why the house burned down on June 11, 2014 rather than some earlier date, then there is an intuitive sense in which her understanding of why the house burned down connects with her knowledge of when it burned down that is not captured by GH3. But this sits quite comfortably with quasi-explanationism, for it suggests the following:

---

[14] I use "knows" here, but nothing would change if this were switched to "has true beliefs about," and so forth.

GH4. Gabe and Holly have identical understanding of why the house burned down, but only Holly understands why the house burned down on June 11, 2014, rather than some earlier date.

Since both pieces of understanding in GH4 are explanatory, nothing here requires the positing of a distinct kind of objectual understanding.

Indeed, precisely because information about when the fire occurred can play a more obvious explanatory role (as in GH4), this may help to account for its lack of goofiness. By contrast, knowing Mandarin or the firefighter's middle name is tangential to understanding of the house fire because neither concerns information that could plausibly play *any* explanatory role in why the house burned down. All of this accords nicely with quasi-explanationism; by contrast, it's unclear how the robust objectualist accounts for their tangential character without adverting to explanatory considerations.

Taking stock, we've seen that when it comes to robust objectualists' reliance on breadth, there is very little wiggle-room between redundancy and excess. Breadth requires that objectual understanding increases as one is able to answer more questions about a subject matter. However, the most obvious questions for which this is true have explanatory import. As a result, these sorts of questions render objectual understanding redundant and thereby fail to establish robust objectualism. So, the robust objectualist can only use breadth to establish her position by finding questions with no explanatory import. However, as we've seen, answers to many of these questions don't seem to be very good candidates for enhancing objectual understanding at all, much less establishing its distinctiveness vis-à-vis explanatory understanding. Rather, they seem to send erstwhile understanders on wild goose-chases – they're excessive.

Now, admittedly, this isn't watertight. Perhaps answers to other questions are both relevant to a subject matter and swing freely of any explanatory considerations. If so, this is a clear place where robust objectualists need to say more than they have: the proof of the pudding is clearly in the eating. Until then, only quasi-explanationism deserves a spot at the table.

4.3. Austere Breadth

Of course, robust objectualists' appeal to breadth is only one of four ways to vindicate objectual understanding's distinctiveness. Perhaps the most underdeveloped way to do so is to justify *austere* objectualism using objectual understanding's (allegedly) greater breadth. Roughly, the idea is that because subject matters aren't limited to explanatory information, one can have understanding of a subject matter while only grasping non-explanatory information.[15]

Unlike updated robust objectualism, updated austere objectualism has a single foil with which it should be compared: minimal explanatory understanding (as defined by EKS2). Moreover, there must be something extraneous about minimal explanatory understanding if updated austere objectualism is to be redeemed. Hence, for the purposes of setting up a fair comparison, I propose the following account of minimal objectual understanding:

> S has minimal understanding of O if and only if for some p about O, S believes that p, and p is approximately true.[16]

The first thing to note about this account of minimal objectual understanding is that it also invites charges of epistemic hoarding. For instance, absent any explanatory considerations, it is hard to see why the firefighter's middle name is not "about" the house fire. Hence, according to the proposed account of minimal objectual understanding, a person with a true belief about this has minimal objectual understanding of the house's burning down. This is implausible to say the least, but I leave it to austere objectualists to rectify this situation.

If we put this difficulty aside, this account of minimal objectual understanding serves as an interesting candidate for austere objectualism, especially when we contrast this with my account of minimal explanatory understanding, which requires belief in an approximately true explanation. Reasonably, any explanation can be cast within a domain, so any case of minimal explanatory understanding would also be a case of minimal objectual understanding. However, the converse seems far less plausible. For instance, consider the following:

> Mona is wholly ignorant of Bahamian brown anole populations, with one exception: her friend, a leading expert on the topic, has recently told her that Bahamian island A13 had no northern curly-tails before or after Hurricane Floyd and had a similarly sized anole population in October 2000 as it did in April 1998. Mona has no beliefs about the causes of brown anole extinction on other Bahamian islands.

---

[15] Kelp's (2015) account of understanding is consistent with this approach, though he does not develop this particular aspect of his view in detail.

[16] If one wanted a tighter connection to Carter and Gordon's conception of breadth, one could stipulate that p is an answer to a question about O. For simplicity's sake, I omit this in what follows.

Quite clearly, Mona has knowledge that pertains to the extinction of some Bahamian islands' brown anole populations in 1999, but lacks a true belief about any explanations thereof. So, given the proposed account of minimal objectual understanding, then Mona has some understanding of Bahamian brown anoles. We can regiment the argument lurking behind this thought experiment as follows:

Austere Breadth Argument

AB1.   Mona has a true belief that Bahamian island A13 had no northern curly-tailed lizards before or after Hurricane Floyd and had a similarly sized anole population in October 2000 as it did in April 1998; i.e. (9) above.

AB2.   This true belief entails that she has minimal objectual understanding of the extinction of some Bahamian brown anole populations.

AB3.   This true belief does not require her to have minimal explanatory understanding of why some Bahamian brown anole populations went extinct.

AB4.   ∴ *Austere Breadth:* Even in the absence of minimal explanatory understanding, objectual understanding can be achieved by having true beliefs about non-explanatory information. (AB1–AB3)

Obviously, both AB1 and AB3 are true *ex hypothesi*. However, AB2 is not something I must grant. In particular, my preferred alternative to AB2 is:

AB2*.   Mona's true belief about (9)'s explanatory role entails that she **is on the right track** to understanding **why** some Bahamian brown anole populations went extinct.

Of course, one can be on the right track without having attained one's intended terminus. Hence, there is no tension between AB2* and AB3. Indeed, as we saw earlier (Section 4.1.2), my account of proto-understanding readily accommodates this proposition about island A13. Specifically, this proposition plays an *indirect* explanatory role as a piece of evidence used to adjudicate between different explanatory models of the anole extinction. In other words, we have no reason to think that AB2 is more plausible than AB2*. Hence, the argument for austere objectualism (at least as AB4 characterizes it) is unsubstantiated.

One may object that because I have gone beyond the strictures of the EKS Model, I have conceded everything that the austere objectualist could wish for. Let's consider this objection in more detail. Recall that explanationism states the following:

Explanationism: Nothing of philosophical importance is lost if all instances
of objectual understanding are treated as instances of explanatory
understanding.

However, if AB2* is true, then Mona only has proto-understanding.
According to the objection under consideration, since proto-understanding is not explanatory understanding, there is a kind of understanding that
evades the EKS Model. While I concede that there is something epistemically laudable here that evades the EKS Model, I deny that it is *understanding*; it is merely *proto*-understanding.

Lest one think my distinction between proto- and minimal understanding-why is arbitrary, recall that I motivated my account of the latter in the
previous chapter. Roughly: it seems obvious that to understand why *p*, one
must possess an approximately true answer to the question, "Why *p*?" My
account of minimal understanding (EKS2) requires this; my account of
proto-understanding (PU) does not.

Furthermore, I do not think that conceding that proto-understanding is
epistemically laudable is much of a concession to *objectualists*. Consider my
preferred answer to the Updated Objectual Question:

Quasi-Explanationism: Nothing of philosophical importance is lost if all
instances of objectual understanding are treated as (one or more)
instances of:
(a)   explanatory understanding, or
(b)   being on the right track to explanatory understanding.

Quasi-explanationism suffices to unseat objectualism. To see this,
observe that there is no obvious reason that the first of the following
sentences is a more apt description than the second:

•   Mona doesn't understand why some Bahamian islands' brown anole
    populations went extinct in 1999, but she has minimal understanding of
    these extinctions.
•   Mona doesn't understand why some Bahamian islands' brown anole
    populations went extinct in 1999, but she's on the right track.

Moreover, the second, a consequence of quasi-explanationism, does
not posit objectual understanding. Hence, it suffices to show that we do
not need to clutter our philosophical cupboards with an additional
concept of objectual understanding and that the core concepts of the
EKS Model – explanation and scientific knowledge – suffice. This surely
runs contrary to objectualism.

Furthermore, proto-understanding suggests a natural progression to our understanding that undermines austere objectualism. When we proto-understand, we grasp only a small piece of a larger explanatory puzzle. The goal is to complete the rest of that puzzle – to grasp a proper explanation. That, of course, fits naturally with the general orientation of the EKS Model. If austere objectualists simply mean that some kinds of objectual understanding are way stations to explanatory understanding, then the debate is merely verbal. Moreover, it seems clear to me that nothing philosophically important hinges on this terminological difference. Hence, I conclude that quasi-explanationism is a suitable foil to austere objectualism – no significant turf has been ceded.

Another challenge to supplanting AB2 with AB2* concerns the fact that, on my view, it's not enough simply to have a true belief in (9); one must also recognize its explanatory role (see Section 4.1.2). Thus, one must also be cognizant of the fact that, e.g., this evidence indicates that Hurricane Floyd alone could not explain the extinction of a Bahamian island's anole population. By contrast, the proposed account of minimal objectual understanding does not require this. Hence, if this proposal is correct, then there is a way of understanding phenomena that falls short of proto-understanding.

However, here I am inclined to simply bite the bullet and reject AB2 if it is so cavalier about explanatory roles. If Mona did not know about the explanatory import of the evidence that she had obtained, then why not just say that she has a true belief and call it a day? In other words, it would be hard to see the purpose of having a concept of objectual understanding if austere objectualists did not inch their way closer to PU. By contrast, proto-understanding plays a valuable role in crediting people with being on the right track to having explanatory knowledge, and what distinguishes them from a merely having true beliefs in propositions that happen to play an explanatory role is that they also *recognize* that role.

Thus, all told, we have seen that justifying objectualism via breadth works for neither its robust nor its austere incarnations. The EKS Model – slightly expanded to include proto-understanding – replicates all that is alleged to be philosophically important about objectual understanding's greater breadth.

4.4. Robust Coherence

Arguably, coherence is the robust objectualist's most heralded X-factor.[17] Indeed, some scholars identify objectual understanding with "holistic

---

[17] I offer further criticisms of the relationship between understanding and coherence in Khalifa (2016).

understanding" and explanatory understanding with "atomistic understanding" (Gordon 2012; Pritchard 2010). Much like their argument for breadth, Carter and Gordon advance a very similar argument (and the same example of Forrest and Emily) for robust objectualism by appealing to objectual understanding's greater *coherence*. As they write:

> We think it is clear that *objectual understanding* – for example, as one attains when one grasps the relevant coherence-making relations between propositions comprising some subject matter – is a particularly valuable epistemic good. . . . Understanding wider subject matters will tend to be more cognitively demanding than understanding narrow subject matters because *more* propositions must be believed and their relations grasped. (Carter and Gordon 2014, 7–8)

This occasions two initial clarifications. First, Carter and Gordon appeal to an inquirer's ability to "grasp" coherence-making relationships. I will bracket the previous chapter's conclusion and take "grasping" to be a theory-neutral placeholder used to denote whatever relationship a person must have to a body of information in order to understand that information. Hence, nothing in this chapter hinges on idiosyncratic assumptions about what "grasping" entails.

Second, I will take these "coherence-making" relationships to include explanatory, probabilistic, and logical relationships. These are the paradigmatic kinds of coherence-making relationships. My arguments would be unaffected if we populated this kingdom of relationships with more exotic denizens, e.g., analogical or model-theoretic relationships, or if we allowed the elements of a subject matter to include non-propositional contents, such as concepts, models, and the like.

With these clarifications in hand, I will argue that coherence fails to vindicate either classic or updated robust objectualism. Beginning with the classics, we see that Carter and Gordon's appeal to coherence suffers from the same sort of maladies that plagued their appeal to breadth:

Classic Coherence Argument

CC1. Emily's objectual understanding requires her to grasp many coherence-making relationships between different statements about a common subject matter.

CC2. Forrest's explanatory understanding does not require him to grasp many coherence-making relationships between different statements *about a common subject matter*.

CRO. ∴ *Classic Robust Objectualism:* Some instances of explanatory understanding are not instances of objectual understanding. (CC1, CC2)

As with the Classic Breadth Argument in Section 4.2.1, the Classic Coherence Argument is invalid, and for precisely the same reason: an unfair comparison. The fact that Forrest's low-grade explanatory understanding doesn't line up with Emma's high-grade objectual understanding can't settle CRO's debts. As before, we need to compare Forrest's low-grade explanatory understanding of why the house burned down with his low-grade objectual understanding of the house's burning down.

With this comparison in place, however, it then becomes unclear what differences in coherence there could be. For instance, suppose that Forrest only has minimal explanatory understanding, as defined by the EKS Model (EKS2). In other words, he only has a true belief that faulty wiring explains why the house burned down. Then Forrest grasps the explanatory (and coherence-making) relationship between the following two statements:

- The house had faulty wiring.
- The house burned down.

Since even minimal explanatory understanding requires the grasp of a coherence-making relation, it follows that any instance of explanatory understanding requires grasping at least one coherence-making relationship. Moreover, if this is all that Forrest understands about the house's catching fire, then his objectual and explanatory understanding are the same. Nor is this limited to minimal explanatory understanding: as Forrest's explanatory understanding improves, the number of explanatory and evidential support relationships he grasps will increase. As a result, it appears that explanatory understanding, objectual understanding, and coherence all increase in lockstep, so there is no clear way that one can diverge from another.

Hence, we should once again update the Classic Objectual Question. Moreover, we should use a similar "testing procedure" as we did for Robust Breadth: imagine two people with the same level of explanatory understanding and then imagine one of them boosting the coherence of that understanding through non-explanatory links; e.g.:

> Initially, Lester and Monica have the exact same understanding of why the house burned down. Later on, only Monica grasps one further coherence-making relationship pertaining to the house fire, while Lester's understanding remains unchanged. Furthermore, Monica's newly grasped coherence-making relationship plays no role in explaining why the house burned down.

The robust objectualist holds that while both Lester and Monica understand why the house burned down to the same degree, Monica (objectually) understands the house's burning down in a way that no degree of understanding could replicate. More precisely:

Robust Coherence Argument

RC1. Objectual understanding can be improved by grasping more coherence-making relationships within a domain.

RC2. Some coherence-making relationships within a domain play no explanatory role.

RC3. The only way to improve explanatory understanding is by playing an explanatory role.

RC4. ∴ *Robust Coherence:* Some instances of objectual understanding involve coherence-making relationships that no degree of explanatory understanding could replicate (RC1–RC3)

As with robust objectualists' appeals to breadth, there is a failure to update; to wit, because of an analogous dilemma of redundancy and excess.

In this context, charges of redundancy amount to claiming that, initial appearances notwithstanding, the added coherence bottoms out in explanatory roles. In other words, the example under consideration fails to substantiate RC2. For instance, in our toy example, this amounts to arguing that Monica's added coherence-making relationship plays an explanatory role. By contrast, changes of excess amount to claiming that insofar as the added coherence plays no explanatory role, it fails to enhance one's understanding, thereby vitiating RC1.

On the redundancy front, it's worth noting that Hempel and Salmon – two paradigmatic proponents of the received view – both think of explanatory understanding in broadly coherentist or holist terms:

> ... all scientific explanation ... seeks to provide a systematic understanding of empirical phenomena by showing that they fit into a nomic nexus. (Hempel 1965, 488)
> ... my suggestion for modification would be to substitute the words 'how they fit into a causal nexus' for 'that they fit into a nomic nexus." (Salmon 1984, 19)

Of course, it is now clear where the Nexus Principle got its name! As already mentioned in Chapter 1, the EKS Model finds commitments to an exclusively nomic or exclusively causal nexus to be needlessly restrictive; we should think more broadly about the *explanatory* nexus. More to the point, it's quite clear that considerations of coherence are not outside of the jurisdiction of explanation; i.e., there are instances of holistic explanatory understanding. Consequently, the robust objectualist still has not moved out of the shadows in which high-grade explanatory understanding can secure all the coherence a robust objectualist could demand. As such, the coherence of objectual understanding appears redundant: quasi-explanationists can also weave these dense webs of cognitively valuable connections into the fabric of understanding.

Indeed, redundancy might well be the sharper horn of our dilemma. As mentioned earlier, the three paradigmatic kinds of "coherence-making" relationships are explanatory, logical, and probabilistic. Of course, insofar as objectual understanding consists of grasping more explanatory relationships, my quasi-explanationism looks rosy. Moreover, there are many accounts of explanation (e.g., causal, deductive, analogical, model-based, unificationist, mechanistic, functional, probabilistic, and intentional-action) and good reasons to think that each of these forms of explanation is permissible in some contexts.[18] Consequently, a wide variety of coherence-making relationships are explanatory and will thereby render many robust objectualist appeals to coherence redundant.

This still leaves logical and probabilistic relationships. Note that some of these will also be explanatory relationships for there are deductive (Friedman 1974; Hempel 1965; Kitcher 1989) and probabilistic (Hempel 1965; Hitchcock 1999; Jeffrey 1969; Railton 1978) models of explanation. Assuming, as I have throughout, that these will sometimes serve as local constraints on specific explanations, there will be cases when the explanatory *relation* is coextensive with a logical or probabilistic relation. However, logical and probabilistic information can also figure in the other direct explanatory roles discussed earlier:

- Logical and probabilistic information may figure in an *explanans*. Railton's DNP model, discussed in Section 4.5.3, is an illustration.
- Furthermore, pieces of logical and probabilistic information can also serve as *explananda*. On the probabilistic side, information about correlations is frequently explained. For example, the fact that smoking causes cancer explains why smoking is correlated with cancer. Similarly, Kepler's laws contain logical relationships (primarily the identity relation) and were explained by Newtonian mechanics.

---

[18] Many of the general reviews of the explanation literature discuss these different forms of explanation, e.g., Cartwright (2004), Lipton (2004 ch. 2), Lycan (2002), Salmon (1989), Thagard (1992, 118–130), Woodward (2002). On the "theory" of explanation I offered in Chapter 1, the variation in these kinds of explanations will result from differences in their "local constraints."

- Logical and probabilistic relations can figure in the *presuppositions* of an explanation (Garfinkel 1981; Hitchcock 1999; Risjord 2000; Sober 1986). Our earlier example, in which certain explanations of brown anole extinction entail commitments about predation (8) is an example.[19]

However, this does not exhaust the explanatory roles that logical and probabilistic information can play. Additionally, such information can play an indirect explanatory role when it does not figure directly in the explanation, but, per the Scientific Knowledge Principle, still plays a role in how scientists come to have explanatory knowledge. In particular, logical and probabilistic relations often indicate the deductive and inductive consequences of a correct explanation. This is why explanations frequently have predictive import (Douglas 2009). For instance, suppose that Monica correctly grasps the following in the course of evaluating potential explanations of why the house burned down:

- That the house had faulty wiring (correctly) explains why it burned down;
- That the breaker box shorted is an alternative, potential explanation of why the house burned down;
- That the house had faulty wiring entails that there would be burn patterns near the outlet in the kitchen;
- That the breaker box shorted entails that there would be no burn patterns near the outlet in the kitchen;
- There are burn patterns near the outlet of the kitchen.

Since I take scientific explanatory knowledge to be the paradigmatic form of understanding, clearly there is room for including non-explanatory relations, such as the logical entailments described in the third and fourth statements, in this capacity. Parallel cases can be made by replacing these entailments with probabilistic relations. Hence, all of this suggests that a good deal of the greater coherence alleged to be objectual understanding's province is illusory; it is redundant given the kind of coherence that explanatory understanding enjoys.

Indeed, the finer-grained details of Carter and Gordon's example admit of similar treatment. First, many of the coherence-making relationships that Emily grasps are naturally construed as playing explanatory

---

[19] Some authors (e.g., Hindriks 2013) have claimed, independently of concerns about objectualism or coherence, that contrastive reasoning provides some evidence of understanding without explanation. As the preceding indicates, I think that contrasts simply play an explanatory role.

roles. For instance, Carter and Gordon claim that Emily's study of the house's wiring provides richer objectual understanding. However, this part of her objectual understanding is naturally seen as providing her with understanding of *how* the wiring caused the fire, and that is just another explanation in addition to understanding *why* the house burned down. Similarly, much of the understanding of both exothermic reactions and stoichiometry that is relevant to the house's burning down explains, e.g., how the fire started/behaved. All of this point to redundancy.

Thus, the EKS Model can incorporate coherence-making relationships in any number of ways: by playing a direct role as an explanatory relationship, explanans, explanandum, or explanatory presupposition; or by playing an indirect role in SEEing. Consequently, any of these relationships can figure in explanatory understanding. Insofar as they do, explanatory understanding is sufficient for capturing anything that's allegedly "special" about coherent, objectual understanding. As a result, the second premise of the Robust Coherence Argument, RC2, will not be satisfied in these sorts of cases. In other words, any coherence-making relationships playing direct or indirect explanatory roles are clearly redundant given the EKS Model. Insofar as only these coherence-making relationships actually enhance explanatory understanding, we should deny updated robust objectualism and opt for updated quasi-explanationism instead.

However, perhaps there are coherence-making relationships that don't play explanatory roles. Could these give updated robust objectualists what they need? Unfortunately for robust objectualists, this launches them squarely into the throes of excess. It's very unclear how explanatorily idle logical and probabilistic relationships could enhance either our objectual or our explanatory understanding. For instance, suppose that Monica is an enthusiast of disjunction-introduction. The fact that she grasps that *the house burned down* entails that *either the house burned down or the moon is made of green cheese* affords her no greater understanding of the house fire than Lester, even though she grasps more logical relationships than he does. Rather, her penchant for or-statements is once again just a perverse kind of "epistemic hoarding." Since such hoarding does not seem to be indicative of rich understanding of a subject matter, *pace* RC1, it is clear that grasping some kinds of coherence-making relationships does not improve objectual understanding.

We can run an analogous argument with Carter and Gordon's description of Emily's richer objectual understanding. Recall that we need fair comparisons. Thus, Carter and Gordon need to show that certain increases in coherence boost Emily's (objectual) understanding of *the house's burning down* without also boosting her (explanatory) understanding of *why the house burned down*. However, such gains in coherence are hard to come by. For instance, Emily's expertise on exothermic reactions is cited as the locus of her objectual understanding, yet many aspects of exothermic chemical reactions are immaterial to the house's burning down. For example, central concepts concerning exothermic reactions (e.g., Gibbs free energy) apply only to closed systems. However, no real-world house fires occur in closed systems. Similarly, certain quantum phenomena are exothermic, but house fires aren't quantum phenomena. Parallel points apply to stoichiometry, the calculation of relative quantities of reactants and products in chemical reactions. Indeed, this also suggests that the *breadth* of Emily's understanding of exothermic reactions and stoichiometry really does not enhance her objectual understanding of the house's burning down, once again casting doubt on RB1. Hence, we have little reason to grant that considerations of coherence show objectual and explanatory understanding to improve in different ways. So, even when we update the Objectual Question, quasi-explanationism is on firm ground.

In summary, I have argued that a fair number of coherence-making relationships can be construed as playing a plethora of explanatory roles. Furthermore, those that can't seem ill disposed to affording us understanding of the phenomenon of interest. Robust objectualists have yet to navigate this dilemma; until they do, coherence doesn't lend credence to their cause.

4.5. Austere Coherence

However, coherence-making relationships have also been used to defend *austere* objectualism. For Kvanvig, understanding has many characteristics that apply to both its objectual and explanatory flavors. Specifically, Kvanvig (2009b, 97) takes both forms of understanding to involve a "grasp of the structural relationships (e.g., logical, probabilistic, and explanatory relationships) between the central items of information regarding which the question of understanding arises."[20] In contrast to explanatory understanding, objectual understanding only "incorporates explanatory relations (*when they exist*)" (Kvanvig 2003, 101: emphasis added). More precisely, we might describe Kvanvig's position as follows:

---

[20] Elsewhere, Kvanvig (2003, 192) describes these explanatory, logical, and probabilistic relationships as "coherence-making." I will alternate freely between "structural" and "coherence" talk.

Kvanvig's Objectualism: S understands *O* if and only if:

KO1.   *S* has several true beliefs about *O*; and

KO2.   *S* grasps several coherence-making relationships that stand between the contents of these beliefs.

Kvanvig's position entails that when explanatory relations do not exist, objectual understanding can still be achieved by grasping other structural relationships (i.e., logical and probabilistic relationships). By contrast, explanatory understanding is ruled out *tout court* in such cases. So, according to this line of reasoning, objectual understanding is broader than explanatory understanding.

However, despite having many similarities with Carter and Gordon's position, Kvanvig puts objectual understanding to very different purposes. After all, he's arguing for *austere* objectualism, and they argued for *robust* objectualism. Given the failures of the latter position, Kvanvig's move might hold promise. To establish this claim, Kvanvig invites us to consider an "indeterministic" system involving, *inter alia*, an electron's trajectory. Kvanvig asserts that while we cannot provide a causal explanation of why the electron went, e.g., to the left rather than the right, we can still have objectual understanding of the system:

> Where *S* is some indeterministic system, we can have objectual understanding of the system even though we cannot interpret this understanding in terms of being able to understand why things happen as they do in *S* . . . So objectual understanding cannot be reduced to propositional understanding via appeal to "wh-"complement attributions of understanding or explanations. (Kvanvig 2009b, 101–102)

Given Kvanvig's account of objectual understanding, this implies that someone grasps logical or probabilistic relationships linking the information that *the electron went left rather than right* to some other information about the system. Since such relationships are presumed to be non-explanatory, it would appear that we (objectually) understand the system, but we do not (explanatorily) understand *why* the electron went left rather than right. With this, we can now present Kvanvig's reasoning:

Indeterministic System Argument

IS1.   *S* grasps many coherence-making relationships among the contents of several true beliefs she has about the quantum system of which the electron is a part.

IS2.   ∴ *S* understands the quantum system (from IS1 and Kvanvig's Objectualism)

IS3.   It is impossible to explain why the electron went left rather than right.

IS4.   If it is impossible to explain why the electron went left rather than right, then it is impossible to understand why the electron went left rather than right.

AC.   ∴ *Austere Coherence:* Even in the absence of minimal explanatory understanding, objectual understanding can be achieved by grasping coherence-making relationships. (IS2–IS4)

Let me begin by highlighting the initial attractions of Kvanvig's position. First, in comparison to other austere objectualist arguments (e.g., Mona's "understanding" of brown anole extinction in Section 4.3), the dearth of explanations about the electron's trajectory isn't simply because of inquirers' lack of imagination. Rather, per IS3, it is widely thought that quantum mechanics entails that certain phenomena, such as an electron's trajectory, *necessarily* defy explanation. Consequently, explanatory under-standing appears to be the wrong concept for the job. Second, and closely related, it seems misplaced to describe someone who understands a system that is *impossible* to explain as having explanatory proto-understanding. Such a person isn't on the right track to arriving at a correct explanation because no such explanation is to be found. Third, because there is still presumably a lot of coherence-making relationships that must be grasped in order to make sense of the quantum system, the kind of bullet-biting I did at the end of Section 4.3 seems far less defensible in this context.

Despite these initial attractions, I will argue that the Indeterministic System Argument provides no sanctuary from quasi-explanationism. The bulk of my efforts on this front will consist of unseating IS3. Specifically, philosophical reflections on indeterministic explanations as they are found in scientific practice strongly suggest that the probabilistic and logical relations involved in Kvanvig's example are actually be explanatory. However, before doing that, I will argue that the Indeterministic System Argument suffers from a more general flaw.

4.5.1.   *Opening Shots*

Recall that austere objectualists claim that anything distinctive about explanatory understanding is merely a contingent feature of a more general notion of objectual understanding. This required them to find a way of showing that there are at least some kinds of objectual understanding that could always be decoupled from explanatory understanding. Austere Coherence states this explicitly:

AC. *Austere Coherence:* Even in the absence of minimal explanatory understanding, objectual understanding can be achieved by grasping coherence-making relationships.

Strictly speaking, this does not follow from the Indeterministic System Argument. The premises of the argument only show us that one can have understanding of a quantum system without understanding why the electron went left rather than right. However, since quantum systems are not the same as electron trajectories, this smacks of an unfair comparison. Indeed, the conclusion that actually follows from this argument is:

IS5. There exists some O such that S understands O and for some p about O, it is impossible to understand why p.

Unfortunately for Kvanvig, IS5 gives us little reason to believe in austere objectualism. To illustrate this point, let me offer two interpretations that are consistent with both IS5 and quasi-explanationism. Here's my first interpretation:

For many p about the quantum system, S understands why p, but S cannot understand why the electron went left rather than right.

In this case, the electron's trajectory is a gap in our understanding of the quantum system, and the rest of our understanding is explanatory. To repeat, this is entirely consistent with IS5 and quasi-explanationism. Indeed, any example that could be run through the Indeterministic Systems Argument (even examples involving deterministic systems!) could be interpreted in analogous fashion, so austere objectualists must take further measures to safeguard against this kind of reply.

For my second interpretation, I'll pull from the stockpile of explanatory roles discussed in earlier sections. Kvanvig's account of objectual understanding rests almost entirely on one's grasp of coherence-making relationships. However, the previous section (Section 4.3) already showed us how to funnel coherentist considerations into quasi-explanationist accounts. Specifically, logical and probabilistic relationships – the "other" coherence-making relationships – can play direct or indirect explanatory roles. Does the electron's trajectory play one of these roles? Kvanvig is silent on this point. Yet it's certainly possible – even if we grant that the electron's trajectory cannot be explained. Consider the following:

Werner understands why the electron's going left rather than right is a non-deterministic process.

For instance, Werner may be well versed in the Uncertainty Principle. Clearly, in this case, that the electron went left rather than right plays a role in Werner's explanatory understanding: it is a presupposition of the explanandum. Hence, someone who only grasped that the electron went left rather than right might be on tap to enjoy explanatory understanding akin to Werner's. In this case, they have proto-understanding of why the electron's going left rather than right is non-deterministic. As argued earlier, my quasi-explanationism readily assimilates instances of proto-understanding.

These alternative interpretations suggest that IS5 is too weak to establish austere objectualism. But even if we bracket this issue, another difficulty looms large. In my two interpretations, I have assumed that no explanation of the electron's trajectory is possible (IS3). However, this is not at all obvious. My primary aim is to highlight the contentiousness of Kvanvig's reasons for taking a broad class of facts – which includes the fact about the electron – to be unexplainable. However, if the electron's trajectory is explainable, then not only can quasi-explanationists fend off the Indeterministic Systems Argument, so too can strict explanationists.

Let's see where Kvanvig's argument hits some snags. Kvanvig takes the conjunction of four general features to preclude the possibility of explaining why the electron went left rather than right (Kvanvig 2009b, 101–102):

- The explanation required is *causal*: "If there is no cause of the electron going left rather than right, there is no explanation why the electron went to the left either."
- The explanandum is *indeterministic*: "In indeterministic systems, things happen that are uncaused, both probabilistically and deterministically."
- The explanandum is *contrastive*: "The events in question are irreducibly indeterministic in such a way that there is no causal explanation as to why the actual events occurred *rather than some other events*" (emphasis added).
- The explanandum contrasts *equally probable* outcomes: "If the probability of an electron going to the left is precisely the same as that of going to the right (and there is no hidden variable to account for the difference), then whichever way it goes is the result of chance rather than causation."

Thus, Kvanvig is denying the possibility of causal, indeterministic explanations of explananda contrasting equally probable outcomes. However, some of these requirements are illicit, and the rest can be satisfied. I will build my case against this position incrementally, addressing each requirement in turn.

4.5.2. Causal Explanation

Let's look at the first requirement. Neither the EKS Model nor even IS3 require explanations to be causal. Any acceptable kind of explanation (causal or otherwise) of the electron's trajectory is an example of explanatory understanding. Moreover, as we've seen, there are many kinds of explanation, each of which is appropriate in different contexts.[21] For instance, the asymptotic explanations featured in Chapter 2 are generally thought to be non-causal (Batterman 2002).

This is important because concepts of causation that work at higher (e.g., macroscopic) levels of physical description do not tidily apply at the level of fundamental physics (Norton 2007), though certain models of non-causal explanation do (Thalos 2002). For instance, since Hempel (1965), many philosophers of science treat certain kinds of theoretical derivations as explanatory, and fundamental physics often trades in just these derivations (Cushing 1991).[22] As such, while the philosophers discussed in this section consider the explanations they examine to be causal, it suffices for my purposes if the logical, probabilistic, and counterfactual dependence[23] relations invoked later are adequate for causal *or* non-causal explanation.

In Kvanvig's example, scientists typically derive probability distributions about an electron's position from its quantum state (i.e., the set of quantum numbers and the eigenfunction that characterize the possible states of the quantum mechanical system of which the electron is a member). As a result, there is not yet a reason to grant IS3.

4.5.3. Indeterministic Explanation

Given that quantum explanations need not be causal, Kvanvig's requirement about indeterminism becomes somewhat moot. While it is a near-tautology that uncaused events cannot be causally explained, uncaused events may very well admit to non-causal explanation. However, Kvanvig's remarks on this topic suggest misgivings about another notion of indeterminism that figures more prominently in the explanation literature. After

---

[21] See Section 1.3.1 for my "theory" of explanation. As I mentioned there, causation is a "local constraint"; i.e., it applies only to some explanations.

[22] Cushing argues for the converse of Kvanvig's claim: in quantum cases, we have explanations without understanding. However, his argument presupposes that causation and visualizability are necessary for understanding. I find this implausible.

[23] For a discussion of counterfactual, noncausal explanation, see Woodward (2003, 220–221).

presenting the core ideas of this latter notion of indeterministic explanation, I address Kvanvig's objection to it.

In the explanation literature, indeterminism simply means that some events are inherently chancy.[24] This need not mean that events are uncaused, but only that the same causes do not always produce the same effect. For example, a coin coming up heads is undoubtedly caused by a toss, but if coin tossing is an indeterminstic process, then an identical toss could have also caused the coin to land tails. Consequently, the core idea of indeterministic *explanation* is that "a factor A is explanatorily relevant to [an explanandum] E if A plays a non-eliminable role in determining the probability of E," where A exhausts the explanatorily relevant information (Hitchcock 1999, 587). Hereafter, I will use "indeterminism" in this sense.

Many prominent theories of explanation of the past thirty years countenance indeterminism (e.g., Humphreys 1989; Lewis 1986; Railton 1978, 1981; Salmon 1984; Woodward 2003).[25] Peter Railton's Deductive Nomological Probabilistic (DNP) Model is one of the earliest and best-known expressions of indeterministic explanation.[26] Formally, Railton presents his model as:

(R1) $\forall t \forall x [F_{x,\, t} \rightarrow \mathrm{Prob}(G)_{x,\, t} = p]$

"At any time, anything that is F has probability p to be G."

Next, we adduce the relevant fact(s) about the case at hand, e:

(R2) $F_{e,\, t0}$

"e is F at time t_o,"

and draw the obvious conclusion:

(R3) $\mathrm{Prob}(G)_{e,\, t0} = p$

"e has probability p to be G at time t_o'"

to which we add parenthetically, and according to how things turn out:

---

[24] For a good review of the literature and the core ideas of indeterministic explanation, see Glymour (2007). Glymour also argues that causal, indeterministic, contrastive explanations of equally probable outcomes are possible, but the details of his view are not necessary for what follows.

[25] Indeed, the only prominent outlier is Kitcher (1989), who denies indeterminism on the grounds that explanations must be deductive arguments. As it turns out, determinism and an explanatory deductivism quite different from Kitcher's might be compatible. Moreover, Kitcher's deductivism has its own host of problems (Glymour 2007).

[26] I use Railton's account only because of its popularity and elegance, but, should another account of explanatory indeterminism prove more satisfactory, my arguments about understanding would remain untouched.

(R4) $(G_{e,\ to}/\neg G_{e,\ to})$

"(e did/did not become G at t_o)." (Railton 1978, 218: my numbering)

Here, the inference from R1 and R2 to R3 must be sound, and the addendum R4 must also be true. Railton also requires that R1 refer to a "causal law," though for our purposes, any legitimate explanatory generalization will do.

Some of Kvanvig's remarks suggest that he is skeptical of this kind of indeterminstic explanation. For instance, Kvanvig (2009b, 101) criticizes those who countenance explanations that "reify chance" into "a further explanatory or causal factor." Indeterminists take the connection between an event's probability to its actually happening to figure in an explanation (e.g., the move from R3 to R4 in Railton's general schema). So, Kvanvig may be rejecting explanatory indeterminism outright.

Admittedly, the explanatory viability of such parenthetic addenda is not immediately intuitive, but there are two responses to Kvanvig's objection. First, indeterministic explanations are staples of scientific practice. For instance, Railton offers the following example:

(1)   All nuclei of ^{238} U have probability $(1 - \exp(-\lambda_{238} \bullet \theta))$ to emit an alpha-particle during any interval of length θ, unless subjected to environmental radiation.

(2) u was a nucleus of ^{238} U at time t_o, and was subjected to no environmental radiation before or during the interval $t_o - (t_o \pm \theta)$.

(3) $\therefore$ u had probability $(1 - \exp(-\lambda_{238} \bullet \theta))$ to emit an alpha-particle during the interval $t_o - (t_o + \theta) \ldots$

(4)   A parenthetic addendum to the effect that u did alpha-decay during the interval $t_o - (t_o + \theta)$. (Railton 1978, 214)

Other examples of indeterministic explanation pepper the natural and social sciences.[27] As a result, if Kvanvig is denying this kind of explanation *tout court*, he is at odds with scientific practice, home to some of our most exemplary feats of explanation and understanding.

Second, Kvanvig's objection does not acknowledge that indeterministic explanations also include theoretical statements like R1 and R2. Scientists do not baldly appeal to a mysterious explanatory factor called "chance" but link a single occurrence – such as an electron's trajectory – to well-defined physical propensities of a system, derivable from the theory (of which the quantum state is a part). Thus, the probabilities or "chances" that Kvanvig

---

[27] Craver (2007) provides some neuroscientific examples; Woodward (2003), examples from a wide variety of sciences.

decries as non-explanatory are derived from theories that are undoubtedly explanatory.

Considerations of the counterfactuals operating in examples of indeterministic explanation highlight the explanatory import of these theoretical elements. For instance, in Railton's example, if u were not a ^{238}U nucleus (but rather, e.g., an ^{241}Am nucleus), the probability of alpha-decay would be different. Since these counterfactuals seem to track closely with our intuitions about explanatory relevance,[28] there is no reason to resist this kind of explanation.

Importantly, a non-contrastive variant of Kvanvig's example lends itself to indeterministic explanation. Derivations from quantum states can tell us why an electron had a probability p of being in a spatial region x at a given time interval t. Explanatory indeterminism then suggests that we can explain why the electron was in x during t simply by treating it as a "parenthetic addendum" to the derivation. As with Railton's example, the counterfactual holds: had the quantum state been different, then the probability of the electron being in a spatial region (e.g., "the left") would be different. Consequently, the indeterministic elements of this explanation do not lend credence to IS3 in and of themselves.

4.5.4. *Contrastive Explanation*

Thus, the orthodoxy on indeterministic explanation puts a good deal of pressure on Kvanvig to capitulate, minimally, that we can explain why an electron went left, even if they would agree with him that we can't explain why it went left rather than right. Indeed, despite the aforementioned authors' commitment to indeterministic non-contrastive explanation, all (save Woodward) think that "Contrastive Explanations Imply Determinism (CEID)" (Hitchcock 1999, 586). Kvanvig (2009b, 101–102) also endorses CEID:

> ... the events in question are irreducibly indeterministic in such a way that there is no ... explanation as to why the actual events occurred rather than some other events.

If correct, CEID might appear to advance Kvanvig's case significantly. I will first examine the justification for CEID and then argue that the justification throws objectualists headlong into a dilemma. Specifically, the argument for

---

[28] Importantly, these theoretical claims allow us to ascertain how the explanandum would have changed if the explanans had been different, an important feature of explanation (Woodward 2003).

CEID strongly favors quasi-explanationism. Alternatively, if CEID is unjustified, quasi-explanationism is still the leading option since we then have an explanation of the electron's trajectory!

So why should we think that CEID is true? The leading argument runs as follows:

P1.　　　If A is a contrastive explanation of E *rather than* F, then F would have been more likely had A not happened.

P2.　　　If F would have been more likely had A not happened, then A does not explain F.

P3.　　　If E and F are descriptions of two different outcomes of an indeterministic system, then A explains why E, and A explains why F.

CEID.　　$\therefore$ If A is a contrastive explanation of E *rather than* F, then E and F are not descriptions of two different outcomes of an indeterministic system.

Let's briefly talk through these premises. P1 captures the idea that contrastive explanation requires an explanans to be a "difference-maker," i.e., to "discriminate" between the outcomes contrasted in an explanandum. For example, when we ask why Adam ate the apple rather than a candy bar, we are looking for something that favors Adam's eating the apple over the candy bar (e.g., Adam was on a diet). Such difference-making is captured by the relevant counterfactual: had Adam not been on a diet, then he would have been more likely to eat a candy bar.

P2 captures the idea that many (if not all) explanations require a relationship of counterfactual dependence. For instance, Adam's diet does not explain why he ate a candy bar. Of course, insofar as my requirement that explanations cite difference-makers tracks with counterfactual dependence, I also grant this point. If Kvanvig does not endorse this premise, then we are owed some story about how he is conceiving of explanation.

Finally, P3 captures what is peculiar about indeterministic explanations. Following Glymour (2007, 139), call this the thesis of *parity*: "one can [indeterministically] explain unlikely outcomes just as well as one can [indeterministically] explain their more probable alternatives." Essentially, parity arises because the same factors produce both a likely outcome and an unlikely one – that is the crux of indeterminism. For example, R4 in Railton's DNP schema appears to clash with contrastivism as it suggests that we should explain e's *becoming G* or e's *not becoming G* in exactly the same way (i.e., by appeal to the derivation of R3 from R1 and

R2). As a result, there is nothing that can differentiate contrasted, inde-terministic outcomes, giving credence to CEID.

Unfortunately for Kvanvig, parity strongly suggests explanationism. If parity holds, then the same information explains all of the possible out-comes of a given system. Thus, the same information (non-contrastively) explains why the electron went left, why it didn't go right, why it could have gone right, and more. However, since this is indeterministic, no additional information could be relevant to the contrast. So, any informa-tion that could figure in objectual understanding already figures in expla-natory understanding.

Thus, Kvanvig faces the same dilemma of redundancy and excess that beleaguer other objectualists: something either plays an explanatory role or it doesn't. If it does, then it's fair game for explanatory understanding, and if it doesn't, then we have very little reason to think that it's relevant to objectual understanding. To illustrate how this dilemma plays out in this particular case, consider a fairly common presupposition in the indetermi-nistic explanation literature:

> A will be said to be explanatorily relevant to E when $P(E \mid A \ \& \ B) \neq P(E \mid B)$.
> (Hitchcock 1999, 587)

Here, B refers to background conditions that are kept fixed in order to rule out spurious correlations. So, the only non-explanatory relationships left to grasp are either probabilistically irrelevant or spurious correlations in which the background conditions B are not held fixed.[29] Prima facie, even someone sympathetic to Kvanvig's position should view these probabilistic relations as implausible (i.e., excessive) bases for objectual understanding.

Thus, if we accept parity, we have an explanation of the electron's going left and also an explanation of its not going right (which, incidentally, share the same explanans), but nothing probabilistically relevant to the electron's going left rather than right can be grasped. Hence the contrast is under-stood neither objectually nor explanatorily, and every explanandum in the vicinity of the contrast (i.e., the system's possible outcomes) is understood explanatorily. Thus, while IS3 would be true, Kvanvig's example would only partake in explanatory understanding. If, on the other hand, parity doesn't hold, then CEID lacks any obvious justification.[30] This paves the way for contrastive indeterministic explanations, such as Hitchcock's:

---

[29] Note that this is an instance of "epistemic hoarding," which also plagued our other objectualists.

[30] For more on parity and/or contrastive indeterministic explanation, see Glymour (2007), Hitchcock (1999), and Strevens (2000).

A is explanatorily relevant to E *rather than* F when $P(E \mid (A \& B) \& (E \vee F)) \neq$ $P(E \mid B \& (E \vee F))$. (Hitchcock 1999, 587)[31]

Returning to Kvanvig's example, since different quantum states (A) can change the probability of the electron's going left (E), even when it is presupposed that the electron went either left or right ($E \vee F$), the electron's going left rather than right is explainable. In this case, IS3 is false because the only understanding in Kvanvig's example is explanatory. Thus, regardless of one's stance on CEID, explanatory understanding appears to carry the day.

4.5.5.   Equally Probable Outcomes

But suppose that Kvanvig denied CEID. His example still might involve non-explanatory understanding if, as he assumes, indeterministic contrastive explanations are possible only when the probabilities of the contrasted outcomes are different. Since Kvanvig's example also assumes that the probability of the electron's going left is equal to that of its going right, indeterministic contrastive explanations might not be possible in this particular case, even if they are possible elsewhere.

Such a position confuses the source of explanatory relevance or difference-making. An explanation needn't make the probabilities between contrasted outcomes different from each other; rather, these probabilities must be different than they would be had the explanans been different.[32] For example, suppose that my being aware of a sea bass dinner special yields a fifty-fifty chance that I choose that special – otherwise I order my old standby, eggplant. However, if I'm unaware of the special, then there is a negligible chance that I order sea bass. Intuitively, my being aware of the special is explanatorily relevant to why I ordered sea bass rather than eggplant.[33]

As this example illustrates, Hitchcock's account of explanation, $P(E \mid (A \& B) \& (E \vee F)) \neq P(E \mid B \& (E \vee F))$, is consistent with equally probable outcomes; i.e., $P(E \mid (A \& B) \& (E \vee F)) = P(F \mid (A \& B) \& (E \vee F))$. Indeed, there is something odd about denying this. Suppose that we moved a polarizer in a continuous fashion and observed the corresponding changes in photon transmission and absorption. On Hitchcock's view, we can explain photon transmission and absorption for every conceivable

---

[31] The disjunction $(E \vee F)$ is exclusive.
[32] While this is consistent with Hitchcock's view, Woodward (2003) provides additional details.
[33] This is a variation on Hitchcock (1999, 602–606).

orientation of the polarizer. By contrast, if there is something special about equal probabilities, the physical connection between the polarizer's orientation and the photon's behavior must be momentarily and miraculously interrupted precisely when the probability of transmission and absorption are identical, but then miraculously resume as soon as those probabilities differ. This is highly counterintuitive, to say the least. Thus, equally probable outcomes readily admit to contrastive explanation.

Taking stock, austere objectualism holds that objectual understanding requires less than explanatory understanding. Kvanvig's Indeterministic System Argument provides an interesting defense of this claim, but it suffers from two flaws. First, its proper conclusion is compatible with quasi-explanationism. Second, it makes false assumptions about explanation. General lessons can be gleaned from this discussion. Despite the fact that the example of electron trajectories challenges some strong folk intuitions about explanation, explanationism emerged unfazed. This suggests that explanations pervade our understanding because causal explanations, deterministic explanations, non-contrastive explanations, and contrastive explanations of events with different probabilities are far less foreign than what we've been considering here. Additionally, the denial of parity is the most contentious move in my discussion, but, as we saw, parity favors quasi-explanationism. Consequently, even if we abstract away from the particulars of Kvanvig's example, austere objectualism may still be very hard to defend.

4.6. Conclusion

My task in this chapter has been to offer an affirmative answer to the question, "Are objectual and explanatory understanding identical in all philosophically important ways?" I have considered the four leading candidates for what would count as a philosophically important difference between objectual and explanatory understanding – breadth and coherence in both their robust and austere flavors – and found that none of them eludes the scope of my account of explanatory understanding, the EKS Model.

Instead of claiming that objectual understanding requires more than explanatory understanding, I have suggested that it is simply an abundance of explanatory understanding. Other times, the objectualist holds up an apple to the explanationist's orange. The corrective for this – what I've been calling "fair comparisons" – frequently results in objectual and explanatory understanding differing only with respect to their surface grammar but otherwise referring to one and the same cognitive achievement. Finally, the most novel revision to the EKS Model has occurred when the objectualist insists that her preferred brand of understanding requires less than explanatory understanding. Here, I have suggested that we need a notion of proto-understanding – of being on the right track to having explanatory understanding. These three scenarios have sufficed to replicate anything alleged to be philosophically distinctive about objectual understanding.

Along the way, I have also highlighted difficulties that arise when objectualists attempt to depart from the EKS Model. In particular, all versions of objectualism run the risk of "epistemic hoarding," of incorrectly prizing the acquisition of trivial or irrelevant information as bona fide advances in our understanding.

So, after this quasi-explanationist polemic, the only differences between objectual and explanatory understanding that survive are the grammatical ones. This difference is not without some utility. We might find it less cumbersome to say, "Tom understands the extinction" rather than "Tom understands why some Bahamian islands' brown anole populations went extinct." Furthermore, given the fact that quasi-explanationism refers to cases where an inquirer might possess a good deal of explanatory understanding, garden-variety explanatory understanding, or mere proto-understanding, objectual locutions might be useful when we're unsure just how much explanatory understanding to attribute to a person. Having said this, these are merely terminological conveniences. Consistent with the larger arc of the book, only explanation and scientific knowledge do the philosophical heavy-lifting.

At this point, it's worth reminding ourselves of the bigger issues at stake. Critics of the received view worry that an epistemology of scientific explanation is too crude to provide a cherished theory of understanding. Thus far, we've seen that my homely little EKS Model handles many facets of understanding – including objectual understanding – more nimbly than those who would seek bolder departures from the received view.

CHAPTER 5

Understanding Without Explanation?

Conventional wisdom suggests that understanding-why is inextricably bound to explanation. Understanding-why requires having an answer to a why-question. Such answers, in turn, are typically identified with explanations. For instance, it borders on the absurd to claim that a doctor understands why her patient has measles while denying that she can identify the cause of those measles – the dreaded rubeola virus.

Despite these attractions, some wish to claim that we can understand why something is the case even when explanations are nowhere to be found. Following Lipton's (2009) posthumous essay, some authors (e.g., Hindriks 2013; Turner 2013) have argued that there can be so-called *understanding without explanation*. Lipton's arguments appear to attack the received view at its core, for the latter follows the conventional wisdom that explanations are necessary for understanding why something is the case. More precisely:

> *S* understands why *p* if and only if there exists some *q* such that *S* knows that *q explains why p*.

In this way, Lipton raises the *Explanation Question*, first mentioned in Chapter 1:

- Does understanding require explanation?

While proponents of the received view answer this affirmatively, Lipton denies this; i.e.:

> *Understanding Without Explanation* (UWE): For some *p*, some people understand why *p*, despite not grasping any correct explanations of *p*.

Confession: I initially found this claim incredulous. Since explanations are answers to why-questions, it would seem that understanding-why just is explanatory understanding. Thus, UWE has the odd-sounding result that there can be explanatory understanding without explanation! However,

Lipton's arguments are characteristically provocative, considered, and worth taking seriously.

Indeed, my initial skepticism notwithstanding, if UWE is correct, not only is the received view false, but so is my preferred successor to it, the Explanation-Knowledge-Science (EKS) Model. Both views make possession of explanatory information necessary for understanding why something is the case. In particular, recall my account of minimal understanding:

> (EKS2) S has minimal understanding of why p if and only if, for some q, S believes that q explains why p, and q explains why p is approximately true.[1]

Hence, if Lipton's arguments are sound, I'm in deep trouble.

Having said this, I think that conventional wisdom still carries the day. Section 5.1 presents and clarifies certain aspects of Lipton's general framework. Section 5.2 then provides three strategies for reconciling UWE with the EKS Model. Sections 5.3 through 5.6 then apply these strategies to Lipton's myriad examples of understanding without explanation. In Section 5.7, I conclude by suggesting that the central premise in Lipton's argument for understanding without explanation is naturally seen as a consequence of the EKS Model. Thus, initial appearances notwithstanding, Lipton's arguments for understanding without explanation presuppose core features of my account.

5.1. Lipton's Framework

Let's begin by presenting Lipton's arguments in some detail. These arguments make two assumptions. First, like me, Lipton is concerned only with cases of understanding *why* something is the case:[2]

> ... my present purpose is to show that, even on a narrow conception of understanding as understanding why, we may nevertheless get understanding without actual explanation. (Lipton 2009, 54)

Lipton's second assumption is that "it is more natural to identify understanding with the cognitive benefits that an explanation provides rather than with the explanation itself" (2009, 43). Call this *Lipton's Assumption:*

---

[1] The other pillar of the EKS Model, EKS1, which accounts for better understanding in terms of the quantity of explanatory information grasped, and the resemblance of that grasp to scientific knowledge, plays a relatively muted role in this chapter, for if I can show that *minimal* understanding requires explanation, *a fortiori*, this applies to more demanding instances of understanding.

[2] More precisely, we both focus on any instances of interrogative understanding that take an explanation-seeking question (whether why-, how-, or something else) about empirical matters as their target.

> If an explanation of p provides a kind of knowledge about p, then that kind of knowledge amounts to understanding why p.

This formulation accords with Lipton's other remarks about understanding as a "cognitive benefit." Specifically, Lipton's most precise enumeration of these cognitive benefits is in terms of "four kinds of knowledge: of causes, of necessity, of possibility, and of unification" (Lipton 2009, 43). Thus, on Lipton's view, in different contexts, an explanation q of p provides knowledge – and hence understanding – in one of the following ways:

- the phenomenon cited by q causes the phenomenon cited by p;
- q entails that p is necessary;
- q entails that p is possible; or
- q unifies p within a broader framework.

Quite plausibly, if q is a correct causal or unifying explanation, then it provides its consumers with knowledge of causes/unification. Furthermore, since explanations frequently entail propositions about what is necessary or possible, I assume that most understanders are deductively competent enough to infer modal knowledge from their explanations.

Two further clarifications are in order. First, it is important that Lipton's Assumption construes understanding as the knowledge provided by *correct* explanations. For instance, conspiracy theories and the machinations of invisible demons can well-nigh explain anything, but presumably they provide no bona fide understanding because such explanations are so wildly incorrect. For the purposes of this chapter, I will follow Lipton in assuming that correct explanations must be approximately true.[3]

Second, I follow Lipton in treating explanations as propositional in character. The leading alternative holds that explanations are things in the world, such as causes. This, however, sits uncomfortably with talk of correct and incorrect explanations, as it is odd to speak of correct and incorrect causes. Of course, propositions *about* causes are perfectly acceptable on the view endorsed here.

As should already be obvious, Lipton's Assumption gives explanation a privileged role. I return to this point in Section 5.7. For now, observe how

---

[3] In Chapter 6, I offer a more nuanced discussion of what explanatory correctness does and does not entail. Introducing those nuances here would needlessly complicate the discussion.

this assumption still sits comfortably with his main thesis – that there can be understanding without explanation:

> The switch from identifying understanding with explanation to identifying it with some of the cognitive benefits of an explanation . . . makes this essay possible. For by distinguishing explanations from the understanding they provide, we make room for the possibility that understanding may also arise in other ways. (Lipton 2009, 44)

Thus, just as burning lumber is not the only way to provide heat, so explanation is not the only way to provide understanding. More precisely, we can reconstruct Lipton's argument as follows:

Lipton's Argument

L1.   *Lipton's Assumption:* If an explanation of p provides a kind of knowledge about p, then that kind of knowledge amounts to understanding why p.

L2.   Explanations provide knowledge of causes, necessity, possibility, or unification.

L3.   Some people have knowledge of causes, necessity, possibility, or unification of certain phenomena but do not grasp any correct explanations of those phenomena.

UWE. ∴ *Understanding Without Explanation:* For some p, some people understand why p, despite not grasping any correct explanation of why p. (L1–L3)

Using the following examples, Lipton devotes the bulk of his efforts to establishing L3:

- Visual models and manipulations can provide tacit knowledge of causes (44–46).
- Non-explanatory deductive inferences can provide knowledge of necessity (46–49).
- Incorrect or "merely potential" explanations can provide knowledge of possibilities (49–52).
- Non-explanatory analogies provide tacit knowledge of unification (52–54).

I will scrutinize these examples later. For now, it suffices to observe that they fit tidily within Lipton's template: a person can obtain the relevant kinds of knowledge by thinking through visual models, manipulations, deductive inferences, merely potential explanations, or analogies without possessing a correct explanation.

5.2. Putting Explanation Back into Understanding

Since Lipton's examples are highly varied, it should come as no surprise that different rebuttals are apt for his diverse cases of understanding without explanation. Here, I provide the general structure of three kinds of rebuttals, one for each of the premises in Lipton's argument. The remainder of this chapter then mixes and matches these dialectical strategies to neutralize Lipton's case for understanding without explanation.

The first strategy challenges Lipton's Assumption (Premise L1 above). Explanations may provide knowledge about a phenomenon, but that knowledge might fall short of understanding *why* that phenomenon exhibits certain properties or behaviors. If we take this route, we can still give these examples a gentle treatment by granting that the examples involve some other cognitive achievement than understanding-why. In particular, the lessons of the previous chapter suggest that some of Lipton's protagonists are "on the right track" to understanding why – what I called *proto-understanding:*

> (PU) S has proto-understanding of why p if and only if:
>   (1) S does not understand why p, and
>   (2) For some proposition q,
>   a. q plays explanatory role R with respect to p, i.e., Rpq, and
>   b. S's mental state toward Rpq is the same kind of mental state toward Rpq that figures in scientific knowledge of why p.

As its first clause indicates, proto-understanding exists without *explanation* but still requires grasping *explanatory roles*. Furthermore, if one grasps a proposition that plays an explanatory role without also grasping the explanation to which that proposition contributes, then, quite clearly, one's understanding looks impoverished when compared to a person who can situate that proposition within a more comprehensive explanatory story.[4] Hence, this understanding without explanation is merely a way station for understanding *with* explanation. Arguably, all of Lipton's examples submit to this *Right Track Objection*.[5]

However, there is room for resisting this objection. In particular, those sympathetic to Lipton's position may claim that it is simply *impossible* for explanations to provide precisely the same kind of understanding as that

---

[4] In this case, one's proto-understanding improves only because one's explanatory understanding has improved. This, of course, fits nicely with the lessons of the previous chapter.
[5] See Sections 5.3, 5.4.1, 5.5.2, and 5.6.2 for arguments to this effect.

afforded by Lipton's non-explanatory pathways. However, note that this undermines the second premise of Lipton's argument (L2): that knowledge of causes, necessity, possibility, or unification are benefits that explanations provide. Call this the *Wrong Benefit Objection*.

At first blush, L2 seems nearly unassailable. However, closer attention to certain examples suggests otherwise. In particular, scientific explanations might well give us understanding of physical necessities, but it is far less clear that they give us knowledge of the kind of conceptual necessity that Lipton may well be presupposing in one of his examples (Section 5.4). Similarly, Lipton may well be relying on a kind of unification that explanations do not afford (Section 5.5). While there might be *some* kind of understanding here, it might not be understanding of *why empirical* phenomena behave in certain ways. Since the EKS Model feigns no hypotheses about non-empirical domains, these examples are unproblematic. They simply highlight that we understand empirical phenomena in different ways than we understand non-empirical matters. Given Chapter 1's disclaimers, this anodyne point poses no threat to the EKS Model.

The third way of rebutting Lipton's argument is to show that, *pace* Lipton, the agents in his examples actually grasp a correct explanation (i.e., to resist L3). A genuine case of understanding without explanation must violate the EKS Model's account of *minimal* understanding:

> (EKS2) *S* has minimal understanding of why *p* if and only if, for some *q*, *S* believes that *q explains why p*, and *q explains why p* is approximately true.

Lipton takes himself to be challenging precisely this claim. However, in Section 3.2, I argued that minimal understanding was quite modest. In some of Lipton's examples (Section 5.5 and Section 5.6), I will argue that this modest kind of understanding is present. In other words, *contra* Lipton, some of his examples involve understanding *with* explanation. Call this the *Explanatory Objection*.

The three rebuttals also crisscross and conspire to put pressure on Lipton's general argumentative strategy. If the understanding could not possibly be achieved via an explanation, then the Wrong Benefit Objection looms large. If, on the other hand, it could be so achieved, then there is yet another tension. Either these examples involve explanations or they do not. If they already drink from the well of explanatory understanding, then the example is actually one of understanding *with* explanation, as the Explanatory Objection states. If, on the other hand, they eschew explanations, then the Right Track Objection rears its head. However, as sketched earlier, the Right Track Objection always leaves proto-understanders with room for explanatory improvement.

Thus, all told, this network of objections suggests that Lipton's examples of understanding without explanation either equivocate on the kind of understanding in play or are otherwise plausibly subsumed under the EKS Model. With this preamble in hand, let's now unleash these objections on Lipton's examples.

5.3. Examples of Modal Understanding

To begin, I will argue that Lipton's claim that "we can gain actual understanding from merely potential explanation" (2009, 49) succumbs to the Right Track Objection. First, let's get clear about how potential explanations instantiate Lipton's argument for UWE. As should be clear, correct explanations provide knowledge of possibilities. Consider my numerous appeals to counterfactual or "what-if" reasoning in characterizing explanations. If we know that q is an explanation of p, then we also know that had q been different, p would also have been different. In other words, we learn of a possibility in which both p and q are different. Hence, it appears that correct explanations provide knowledge of possibilities (L2). For Lipton, not only correct explanations furnish knowledge of what is possible, as "a merely possible explanation may also give information about the modal status of an explanation" (Lipton 2009, 51). Quite clearly, one may only grasp a potential explanation without also grasping an actual (i.e., correct) explanation and thereby gain knowledge of possibilities (L3). Hence, given Lipton's Assumption (L1), Lipton has offered a putative example of understanding without explanation. In other words, Lipton is claiming that we sometimes understand a phenomenon by knowing how it could have been without knowing the actual explanation underlying it.

Lipton provides two clear rationales for how modal understanding without actual explanation is possible.[6] First, potential explanations "may show a degree of contingency in the actual explanation" (Lipton 2009, 51). For instance, suppose that a firm hires Jones because he had extensive prior experience in the industry. Moreover, the contingency of

---

[6] Lipton (2009, 49–52) lists several other examples under this heading, but their rationale is not as explicit as the ones discussed here. I submit that they can be assimilated to at least one of the three strategies presented earlier.

his hiring is highlighted by the fact that his other credentials were fairly nondescript, such that he would not have been hired had he lacked this experience. Thus, the falsehood of other, potential explanations of Jones's hiring highlights its contingency. Suppose that an agent knows that if other hiring criteria (e.g., education) had been the deciding factor, Jones would not have been hired, but does not know of Jones' superior experience. In this case, Lipton claims that the agent possesses some understanding of why Jones was hired, although she lacks knowledge of a correct explanation of Jones's hiring.

Second, potential explanations "may show necessity . . . by revealing fail-safe overdetermination," as the following illustrates:

> Suppose that a boxing match between Able and Baker is rigged so that Baker – though in fact the far better boxer – would take a dive in the tenth round. Knowing this helps us to understand why Able won, even if as a matter of fact Able floors Baker with a lucky uppercut in the fifth. (Lipton 2009, 51)

Suppose that someone knows that if the fight had lasted until the tenth, Baker would have taken the dive, but does not know that Able's fifth-round knockout actually caused his victory. Because this person possesses knowledge of salient counterfactual scenarios, she understands without knowing an actual explanation.

Let's now apply the Right Track Objection to these examples. This rebuttal works in two steps. First, I'll argue that certain cognitive benefits provided by explanations of a phenomenon do not amount to understanding why that phenomenon behaves as it does, but instead garner only some proto-understanding of the phenomenon. Second, I'll show how this proto-understanding would improve if it were upgraded to explanatory understanding.

In Lipton's first example, suppose that someone knows that Jones' education alone would not have resulted in his hiring, but does not know what actually prompted his hiring. When asked why Jones was hired, the answer will have to be indirect or oblique (e.g., "Not because of his education"). By contrast, a person with a correct explanation will be able to state that Jones was hired not because of his education, but because of his experience. Intuitively, the latter person better understands Jones's hiring. Indeed, it's easy to see why: in addition to having knowledge of the contingency of Jones's hiring, she also knows the cause of Jones's hiring. Even by Lipton's own lights, causal knowledge is a mark of understanding. Hence, the explainer has accrued more cognitive benefits provided by an explanation – more *understanding* according to Lipton's own standards – than her explanatorily benighted counterpart.

Similarly, a person who only knew that Baker *could have* lost the match because he would have taken a dive in the tenth, but does not know that Baker was *actually* knocked out fair and square in the fifth, doesn't understand why Baker lost the match. Once again, think of how such a person would respond to the relevant question, viz. "Why did Baker lose the match?" Here, it would be generous to even call the answer indirect: "Well, he would have taken a dive in the tenth." By contrast, someone who also knows about the actual course of events can answer this question directly: because of the lucky uppercut in the fifth. Thus, in both examples, far less understanding is to be had without an explanation.

Let's tighten up the claim that proto-understanding is in play here. Recall that someone who fails to understand why might still be on the right track to understanding why if she grasps the explanatory roles that certain propositions play. As discussed in earlier chapters, explanantia, explananda, descriptions of explanatory relations, and explanatory presuppositions play *direct* explanatory roles. Propositions play *indirect* explanatory roles if they are not direct but still figure in scientific knowledge of an explanation.

This is precisely what we see in these examples: agents gain understanding by knowing what would happen had things turned out differently. As I argued in Chapters 1 and 3, I follow Woodward (2003, 221) in claiming that explanations must answer "what-if questions." Hence, answers to what-if questions are consequences of correct explanations, some of which are essential to scientists' evaluations of which explanations are correct. Thus, the kind of modal knowledge operant in these examples partakes of indirect explanatory roles.[7]

The other part of the Right Track Objection shows that wherever one grasps a proposition's explanatory role without also grasping the explanation in which that role is played, room for improvement is always available. This is where the Right Track Objection gets its teeth: given the choice between answering what-if questions about *p* and answering why *p*, the best option is both!

Arguably, against the Right Track Objection, one might retort that explanatory knowledge does not require knowledge of these *specific* possibilities in question, namely:

---

[7] See Section 4.2.2 for why I take answers to what-if questions to play indirect explanatory roles.

- Had Jones had fewer years of industry experience, he would not have been hired; and
- Had Able not landed the lucky uppercut in the fifth, he still would have won the boxing match.

While it is certainly true that not *all* of our explanations answer every what-if question, the Right Track Objection does not hinge on this point. Rather, it hinges on the fact that it is odd to claim that one understands why *p* when one is unable to provide a correct answer to the question, "Why *p*?" and that there is at least *one* such answer – i.e., one correct explanation – that also answers the same what-if questions as a merely potential explanation.[8] Furthermore, since the ability to answer *more* what-if questions tracks with explanatory goodness, the better one's explanatory evaluation, the more what-if questions one can answer. Consequently, the Right Track Objection survives this challenge.

Let me conclude this section by highlighting the Right Track Objection's generality. Lipton's Assumption implies that any knowledge provided by an explanation is a kind understanding. All direct explanatory roles and the most important indirect explanatory roles (answers to what-if questions) are consequences of explanations. While Lipton does not provide any precise specification of how explanations "provide" knowledge, it is hard to see a more innocuous mode of provision than that of inferring an explanation's logical consequences. Hence, explanations provide knowledge of their logical consequences. Whenever such knowledge falls short of explanatory knowledge, we will have (proto-) understanding without explanation.

However, this always means that one's understanding improves in proportion to the amount of knowledge one has of direct explanatory roles – including knowledge of an entire explanatory nexus! So, there will always be an instance of explanatory understanding that outstrips the proto-understanding being touted as an instance of UWE. For instance, since actual explanations are clearly stronger than their consequences concerning possibilities (e.g., *if q then p is possible*), this means that understanding via actual explanations will always feature claims that cannot be replicated by merely potential explanations. Thus, Lipton's discussion of the understanding afforded by potential explanations – and indeed knowledge of any explanatory consequence whatsoever – does

---

[8] Parallel objections and replies could be rehearsed in each of the following sections. I spare the reader the redundancy.

nothing to challenge explanatory knowledge's status as the bedrock of understanding.

5.4.   The Galileo Example

Let's consider another of Lipton's examples, concerning Galileo's demonstration of why gravitational acceleration is independent of mass. Galileo supposes the contrary and then considers what would happen if a heavier mass m_1 were attached to a lighter mass m_2. Since, *ex hypothesi*, m_2 falls at a lesser speed than m_1, the two masses should fall more slowly than m_1 alone. However, when considered as one mass, $m_1 + m_2$ is heavier and should thus fall faster than m_1 alone. But since the same thing cannot fall both more quickly and more slowly than m_1, acceleration must be independent of mass (Galilei 1914, 62–63; Lipton 2009, 47).

Lipton (2009, 48) contends that Galileo's argument is not an explanation, because:

> Rather than saying directly why acceleration must be independent of mass, the argument works by showing that the contrary assumption would entail a contradiction.

Then, according to Lipton, despite lacking an explanation, Galileo knows that the conclusion of his argument is necessarily true (L3) and, since knowledge of necessity is a cognitive benefit provided by an explanation (L2), it follows from Lipton's Assumption (L1) that Galileo understands why acceleration must be independent of mass. Thus, if Lipton's argument is sound, we have a genuine case of understanding without explanation.

Lipton faces a dilemma here. Either the knowledge of necessity in this example can play some explanatory role or it cannot. If it can, then we can revisit the Right Track Objection (Section 5.4.1). If it cannot, then Lipton faces the Wrong Benefit Objection (Section 5.4.2). Let's examine each of these objections in turn.

5.4.1.   *Critical Information and Proto-Understanding*

The most natural way to finesse the first horn of this dilemma pays closer attention to the argumentative context in which Galileo actually offers his *reductio*. Galileo is criticizing a potential explanation – a point Lipton does not acknowledge. Even a cursory read of the relevant sections of Galileo's *Dialogue Concerning Two New Sciences* shows that his reasoning is a refutation of Aristotelian explanations of motion. Consider Galileo's preamble to his thought experiment:

> *Salviati*: But, even without further experiment, it is possible to prove clearly, by means of a short and conclusive argument, that a heavier body does not move more rapidly than a lighter one provided both bodies are of the same material and in short such as those mentioned by Aristotle. (Galilei 1914, 62)

The character of Salviati is Galileo's Copernican interlocutor in the *Dialogue*. He is responding to Simplicius – the mouthpiece for Aristotelian (and Ptolemaic) ideas. Indeed, a key reason for presenting these ideas in the form of a dialogue is to criticize the Aristotelian and Ptolemaic system.

When we put Galileo in context, it's clear that he is assessing whether or not acceleration *depends* on mass, which is tantamount to considering whether or not an object's being of a certain mass *explains* how it accelerates. So, Galileo treats mass as potentially explaining acceleration. If Aristotelian physics were correct, then m_1's mass would explain its acceleration. But Galileo then considers what would happen if we attached m_2 to m_1, and his *reductio* is a criticism of the Aristotelian explanation; it shows that differences in mass *fail* to explain differences in speed.

However, seen in this light, Galileo is really no different from the protagonists of the previous section. If armed only with his clever thought experiment, a person who is asked, "Why do objects accelerate?" can only offer an indirect answer: "Not because of their mass." Thus, at best, such a person doesn't understand why objects accelerate but is "on the right track."

Recall that my account of proto-understanding requires inquirers to grasp propositions that play explanatory roles. However, whereas the previous section focused on consequences of a correct explanation, the Galileo example trades in other propositions that figure in scientific explanatory evaluation (SEEing). On the interpretation of Galileo being offered here, he has considered a potential explanation of acceleration and has ruled it out through scientific reasoning. However, we've already encountered places where potential explanations played this kind of role in the EKS Model. As discussed in previous chapters, SEEing involves considering and comparing plausible potential explanations. So, Galileo is engaging in reasoning characteristic of an expert explanatory evaluator.

To feed these ideas into the Right Track Objection, let *critical information* be justified true beliefs that potential explanations are incorrect. Thus, Galileo's demonstration provides critical information about how objects accelerate. But, crucially, such information figures in SEEing – specifically at the stage of comparison. So, critical information plays an indirect explanatory role. As such, the EKS Model takes the gathering of critical information as a means of improving one's understanding. Thus, if someone has significant understanding of why there are certain differences in speed, then she is liable to possess this critical information. Furthermore, she will also grasp the correct explanation (e.g., that acceleration is a result of the net forces acting on an object), and this would provide an added cognitive benefit (e.g., knowledge of the causes of acceleration).

Let me anticipate two objections to this argument. First, one might complain that I've pulled a bait and switch: Lipton claims that knowledge of necessity is a form of understanding, and I have changed the topic by rendering critical information the relevant form of understanding. I do not think a close reading of the *Dialogue* supports the accusation that *I've* pulled the bait and switch. Rather, I think it's Lipton who has disregarded the context in which Galileo delivered his thought experiment.

Moreover, Galileo's mastery of critical information appears more fundamental to his understanding than his knowledge of necessity: consider that even if Galileo failed to show that the mass explanation is *necessarily* false, he still would have offered an especially strong criticism. The shift in emphasis from knowledge of necessity to critical information gains further credence when considering arguments denying that Galileo's reasoning amounts to a *reductio* (Gendler 1998; Schrenk 2004; Vickers 2013). In effect, these arguments suggest that Galileo's reasoning only shows the implausibility, but not the impossibility, of the mass explanation. Such an interpretation accords well with my privileging of criticism over necessity.

A second objection is that appealing to critical information only works for understanding obtained via *reductio* but not for Lipton's more general claim that knowledge of necessity via deduction can furnish examples of understanding without explanation. Absent a compelling example, this concern is difficult to address. On the one hand, some deductive inferences are explanatory, so these cannot be used as examples of understanding without explanation. On the other hand, some knowledge of necessity (e.g., knowledge of tautologies) seems to fall short of understanding. While navigating these two extremes is *prima facie* possible, it behooves the objector to offer an example.

Moving back to the bigger picture, we have seen that Galileo's understanding without explanation is simply capturing an aspect of scientific explanatory evaluation: the consideration and evaluation of plausible potential explanations of a phenomenon of interest. Since I regard this as the paradigmatic way of garnering scientific knowledge of an explanation,

Lipton's use of Galileo does nothing to undercut scientific knowledge of the explanatory nexus as the ideal of understanding.

5.4.2. *Have We Misunderstood Galileo's Contribution?*

In conversation, many people still charge me with baiting and switching on this example. While my preferred treatment of the Galileo example is the one just presented, let me grant – but only for the sake of argument – that the Right Track Objection doesn't work and that Galileo's demonstration deserves more credit than "Close, but no cigar." This simply sets the stage for the other horn of our dilemma, in which the Wrong Benefit Objection makes it mark.

At first blush, it may seem like good news for fans of UWE that I've granted that no explanation could replicate Galileo's achievements. However, things are not that simple. After all, Lipton's Assumption holds that understanding is a cognitive benefit that an explanation could provide, and he provides no other criteria for identifying understanding. Hence, once we assume that no explanation of acceleration's independence from mass can provide the same understanding as Galileo's *reductio*, Lipton's argumentative strategy no longer supports the claim that this demonstration provides understanding of acceleration's independence of mass.

Perhaps fans of UWE think that we can bypass Lipton's argument at this point, for it's very intuitive to grant that Galileo's ingenious proof advanced our understanding – to wit, our understanding of *why* acceleration is independent of mass. Indeed, I once shared these intuitions. However, I now think that they can be explained away. Here's my diagnosis: if you don't accept my first interpretation, in which Galileo traffics in critical information, then your intuition rests on conflating the following two claims:

(1) Galileo's proof advances our understanding of the *concepts* of mass, acceleration, and their interrelation.
(2) Galileo's proof advances our understanding of the *empirical phenomena* of mass, acceleration, and their interrelation.

Given the assumption that I've granted for the sake of argument – most notably that no explanation could provide just the understanding that Galileo's thought experiment provides – I think that only the first of these claims is defensible. Note that we are not talking about a quantum phenomenon, where something like a no-hidden-variables proof might well favor a deep kind of *physical* impossibility. The impossibility has to be of a *conceptual* variety. Moreover, Galileo does not require one iota of empirical evidence to push through his demonstration. It is as *a priori* a demonstration as one is likely to see in science.[9] This is one sign that we are not dealing with garden-variety understanding-why.

However, here's the real kicker: we have no strong reason to think that scientific explanations provide *a priori* knowledge of *conceptual* necessities. Hence, the Wrong Benefit Objection surfaces. While there is plenty to be said about how we understand concepts (e.g., Peacocke 2008), for reasons already discussed, this is not within the EKS Model's purview. To use Galileo's thought experiment as a counterexample to the EKS Model is no different than complaining about the EKS Model's "inadequacy" in accounting for our understanding of Arabic or of artworks.

So, to summarize, the understanding we get from derivations such as Galileo's is either achievable via explanation or it is not. If knowledge of necessity plays an explanatory role, then Galileo has proto-understanding that would be improved by obtaining an explanation. This is entirely congenial to the EKS Model. If knowledge of necessity does not play an explanatory role, then Galileo has understanding of concepts, but not of empirical phenomena. This is entirely orthogonal to the EKS Model's intended domain of applicability.

5.5. Unification via Tacit Analogy

Thus far, I have argued that critical information and knowledge of possibilities are cognitive benefits that either explanations or their alternatives can provide. I've then shown that if we only had non-explanatory understanding, we wouldn't have an answer to the why-question of interest. Consequently, the "obliqueness" of the grasped information involved in examples of UWE has thus far proven essential to highlighting the greater understanding provided by explanations.

However, Lipton has other examples of understanding without explanation where the agents might well be thought to *have* an answer to the relevant why-question even if they cannot *give* this answer. In these

---

[9] Return to the earlier quote from the *Dialogue* and observe Galileo's insistence that the understanding in question can be achieved "without further experiment, it is possible to prove clearly, by means of a short and conclusive argument."

examples, Lipton pries apart understanding from explanation by appealing to tacit knowledge.[10] For instance, he writes:

> Kuhn [offers] an account of . . . the scientists' ability to select problems and generate and evaluate solutions. And these abilities correspond also to a knowledge that goes beyond the explicit content of the theory. The exemplars provide knowledge of how different phenomena fit together. The unarticulated similarity relations that the exemplars support provide a taxonomy that gives information about the structure of the world. They thus have the effect of unifying the phenomena, and they do this by analogy, not by explanation. (Lipton 2009, 53)

This suggests that unification without explanation works as follows:

Analogical Unification

- The agent knows the following exemplar: B is a solution to problem A.
- The agent is also able to identify that C is a problem analogous to A and that C has a solution, D, analogous to B.
- So the agent knows that A, B, C, and D are unified.

Since an explanation could provide knowledge of unification (L2), Lipton's Assumption (L1) treats this as a form of understanding. Analogical unification also appears to forgo explanatory knowledge (L3) and is presumably tacit because the person cannot say how C and D are analogous to A and B, respectively. Thus, as before, Lipton alleges that we have understanding without explanation.

This particular example is (in principle) fodder for all three of my rebuttals. I will first show how the Explanatory Objection poses problems for understanding via analogy. This requires carefully distinguishing explanatory unification from "Kuhnian" unification by tacit analogy (Section 5.5.1). However, even if this objection to Lipton's example doesn't work, it still succumbs to the remaining objections (Section 5.5.2).

5.5.1.   Unification with Explanation

Lipton is not explicit about how explanations provide knowledge of unification. However, navigating these issues requires a clearer account of how unification via explanation operates. I'll proceed in two steps. First, recall my general constraints on explanation from earlier chapters:

q (correctly) explains why p if and only if:

(1) p is (approximately) true;

(2) q makes a difference to p;

---

[10] Lipton does not reference any works on tacit knowledge (e.g., Polanyi 2009). It is unclear whether he intends anything precise.

(3)   *q* satisfies your ontological requirements (so long as they are reasonable); and

(4)   *q* satisfies the appropriate local constraints.

Remember that, for the purposes of this chapter, I'm simplifying the third condition so that the explanans *q* must be approximately true.

The second step consists of specifying the local constraints that distinguish unifying explanations from other kinds of explanations. To that end, I'll provide a brief sketch of Philip Kitcher's (1989) influential "unificationist" account of explanation. Essential for our purposes is Kitcher's (1989: 430) claim that "Science supplies us with explanations whose worth cannot be appreciated by considering them one-by-one but only by seeing how they form part of a systematic picture of the order of nature."

More precisely, Kitcher holds that we gain knowledge that an explanation *q unifies p* (i.e., we "see" this "systematic picture") when we know that the derivation of *p* from *q* is an instance of a broader derivation pattern (or schema) that can be applied to phenomena other than *p*. This suggests the following explanatory alternative to analogical unification:

Explanatory Unification

- The agent knows that *B* explains *A* and that *D* explains *C*.[11]
- The agent also knows that *B*'s explaining *A* and *D*'s explaining *C* are both instances of a more general pattern or schema *G*.
- So the agent knows that *G* unifies *A, B, C,* and *D*.

Schemas such as *G* contain information such as the following:

External Pathway Explanation Schema:
Explanation target:
Why does a **cell** become defective in a **function**?
Explanation pattern:
The **cell** is destructively affected by **external agents**, such as bacteria, viruses, or autoimmune cells.
These **external agents** operate by means of **pathways** that enable them to invade and disrupt the **cell**.
So the **cell** becomes defective and cannot carry out its **function**. (Thagard 2003, 244–245)[12]

---

[11]   Many unificationists reduce the explanatory relationship to an inferential relationship and, often more narrowly, a deductive relationship (Friedman 1974; Kitcher 1989; Schurz and Lambert 1994). Changing the word "explains" to "entails" would not affect the points about understanding we make herein.

[12]   Kitcher (1989: 438–448) deploys schemas to depict the progressive unification of evolutionary biology and chemistry. Thagard's (2003) example illustrates Kitcher's idea more compactly.

Here, the boldfaced letters are variables that are filled in by different values. The unifying power of the schema is directly proportional to the variety of values that these variables can assume while still yielding correct instances of the explanation schema. Thus, the local constraints in question will include schemas and will fill those schemas out with specific values that line up with the explanans and the explanandum. Unification is achieved when multiple instantiations of the schema satisfy my three other "global" constraints on explanation.

Importantly, on Kitcher's view, an explanation's providing knowledge of unification *entails* that it does so by way of fitting our explananda into a more general pattern or schema. Since this idea is paramount in what follows, it is worth stressing that it animates most other accounts of explanatory unification. For instance, Friedman (1974: 19) writes that when we explain via unification,

> We replace one phenomenon with a *more comprehensive* phenomenon, and thereby effect a reduction in the total number of accepted phenomena. We thus genuinely increase our understanding of the world.

Bartelborth's (2002: 91) more recent unificationism claims that "explanations promote our understanding of the world [by] embedding ... our observations, events, and other facts into more general patterns that bring our different observations together in a coherent world view." Indeed, even advocates of non-unificationist accounts of explanation often favor explanations involving broader or more invariant explanatory generalizations (e.g., Woodward 2003).

Thus, according to many sustained reflections on this topic, explanations provide this kind of understanding via general schemas or patterns of reasoning. Hence I assume that if explanations provide knowledge of unification, then they provide knowledge of general schemas. Should other accounts of explanatory unification bypass knowledge of general schemas, we would have to evaluate those accounts and their ramifications for the understanding without explanation.

So, we can provide a quick and dirty version of the Explanatory Objection, as we now have clear accounts of how both tacit analogy and explanation provide unification. Suppose that both involve knowledge of a general schema (*G* above). Then Lipton's example is not a genuine case of understanding *without explanation*, as this would amount to the curious claim that understanding without explanation amounts to tacit knowledge of an *explanatory* schema. Indeed, unificationists such as Kitcher, Friedman, and Bartelborth do not deny that these schemas can be tacit.

5.5.2.   *Unification Without Explanation*

In order to avoid the Explanatory Objection, tacit analogies must not involve knowledge of the schemas. As it turns out, such an assumption leads either to the Wrong Benefit Objection or to the Right Track Objection. Let's examine each objection in turn.

Assume, for the sake of argument, that a person can have knowledge of unification via tacit analogy without wielding an explanation. On Lipton's reading of Kuhn, the analogies concern problems and their solutions. Presumably, however, some problems and their solutions have no clear explanatory upshot since they concern other corners of scientific practice. In these cases, it is unclear how the kind of unification afforded by tacit analogy could be a cognitive benefit provided by an explanation. Thus, absent further details, it's unclear why we should attribute understanding-*why* in such cases. Perhaps the agent merely understands *how* to solve certain problems, and Lipton is conflating a kind of practical or procedural understanding with understanding-why. In effect, this is a version of the Wrong Benefit Objection.

So, the problems and solutions must have some explanatory payoff – they must play some explanatory role. But this concession leads directly to the Right Track Objection: grasping any proposition's explanatory role contributes to proto-understanding. And, as we've already seen, once we grasp an explanatory role, our understanding improves by situating that role within a larger ensemble of explanatory commitments – by going further and achieving "full-blown" explanatory understanding.

Let's pinpoint the exact contribution of unifying explanations. In cases of "tacit unification," agents only know *that* similar problems admit of similar solutions, but they cannot provide reasons for *why* this is so. General schemas provide those reasons, and inferential knowledge of this sort is cognitively beneficial. Hence, unifying explanations provide inferential knowledge. Recall that Lipton's Assumption identifies understanding with knowledge afforded by explanations. Hence, by Lipton's own standards, the inferential knowledge provided by explanations is a kind of understanding. To this, we can add the following: understanding improves as we accrue more of these cognitive benefits. While a person with tacit unification and a person who has a unifying explanation both know that certain phenomena and hypotheses are similar, only the latter also knows why this is so. Thus, understanding with unifying explanations is superior to its tacit alternative, just as the Right Track Objection claims.

One might resist this argument by prying apart knowledge of an explanation and knowledge of a schema, but then we are owed some account of how explanations provide knowledge of unification. Moreover, generalizations of some sort (laws, invariant causal generalizations, models that can accommodate other phenomena, etc.) clearly are essential to many explanations, and these allow us to explain multiple phenomena using the same conceptual resources. Thus, regardless of whether *schemas* are the best way to capture this intuition, explanations provide knowledge of unification by fitting an explanandum into a more general framework, and knowledge of this framework differentiates explanatory unification from Lipton's Kuhnian alternative.

Thus, explanations provide unification through knowledge of general patterns. Lipton's example of unification via tacit analogy then suffers a dilemma. Either tacit analogies invoke these patterns and hence do not provide understanding without explanation, or these analogies proceed without those patterns, in which case explanations are superior to analogies precisely because they provide knowledge of these general patterns. All of this bodes well for the EKS Model.

5.6.   Tacit Understanding of Causes

Finally, Lipton uses tacit knowledge to provide two further putative examples of understanding without explanation. First, Lipton describes a person who gains tacit understanding of the causes of retrograde motion through a visual model of the solar system (an orrery):

> These visual devices convey causal information without recourse to an explanation. And people who gain understanding in this way may not be left in a position to formulate an explanation that captures the same information. Yet their understanding is real. (Lipton 2009, 45)

In addition to visual models, Lipton takes manipulation as another site of tacit understanding without explanation:

> . . . a scientist may gain a sophisticated understanding of the behavior of a complicated piece of machinery by becoming an expert at using it, and that understanding consists in part in the acquisition of causal information that the scientist may be in no position to articulate. (Lipton 2009, 45)

In both examples, the inquirers possess causal knowledge, and, per Lipton's Assumption, they thereby possess understanding (L1 and L2). So, why exactly does Lipton think that they lack an explanation? According to

Lipton (2009, 45), this is because explanation "requires that the information be given an explicit representation. In short, there is such a thing as tacit understanding, but not tacit explanation, and this provides the space we are looking for, where there can be understanding without explanation."

Thus, Lipton argues for L3 by assuming that explanations only provide explicit causal knowledge and then showing that some visual models and manipulations provide causal knowledge that is not explicit. Since causal knowledge is a kind of understanding, these models and manipulations are alleged to provide understanding without explanation.

This requires a more precise account of explicit and tacit knowledge. So, let us say that agents who explicitly know that *A causes B* (a) know that *A causes B* and (b) can verbally communicate that *A causes B*. Tacit knowledge differs from explicit knowledge with respect to the second condition (b). In other words, tacit knowers lack the ability to communicate their knowledge verbally. While Lipton does not present many details about how he distinguishes explicit and tacit understanding, his few remarks are congenial to this gloss. For instance, he describes his tacit knowers as being "inarticulate" and unable to "say something" (Lipton 2009, 45).

As before, I pose a dilemma. On the one hand, if grasping an explanation does not require verbal abilities, then the Explanatory Objection kicks in: tacit explanations exist, so these aren't examples of understanding without explanation (Section 5.6.1). Alternatively, if grasping an explanation involves verbal abilities, then the Right Track Objection applies: explainers have a kind of "semantic knowledge" that their counterparts lack (Section 5.6.2).

5.6.1. *Tacit Explanations*

Pace Lipton, I believe that explanations in the absence of verbal communication are possible. If someone has an accurate representation of a difference-making relationship, then, give or take some local constraints, one grasps an explanation. None of this requires the representation to be linguistic. As a result, the Explanatory Objection threatens Lipton's example, and the tacit–explicit distinction should be handled with greater caution when trying to construct examples of understanding without explanation.

I want to push on Lipton's example in three different ways. First, and quickly, I simply think that verbal capacities are not necessary to possess explanatory information. Moreover, a quick review of the explanation literature shows that few philosophers of science have required anything resembling an "explicitness" condition on explanation. However, perhaps this speaks more to certain taken-for-granted assumptions than anything else.

So, let's move to my second way of resisting Lipton's claims: by highlighting the arbitrariness of the tacit–explicit distinction. Imagine a person, Fred, who knows that pushing a lever causes a wheel to spin. However, Fred is mute and expresses this causal claim in sign language. Fred lacks verbal abilities, but he's clearly offering an explanation. In other words, our choice to use verbal symbols over other (e.g., gestural or pictorial) symbols is entirely a matter of convention, and no deep facts about explanation hang on this choice alone. For instance, Fred's expression in sign language translates into a readily identifiable explanation in English, yet it didn't undergo some magical transformation in the process of translation – it was already expressing an explanation in sign language. This is just to say that explanations are fundamentally propositional rather than sentential creatures, in which case their syntactical expressions are largely incidental.

This leaves far less room for non-explanatory tacit understanding. Indeed, any symbol system – verbal or otherwise – with interpretive conventions adequate for the purposes of communicating causal information appears expressively adequate for the purposes of explanation. For instance, suppose that another person, Dan, can use Fred's machine in just the tacit manner that Lipton describes. Suppose further that Dan can push the lever to demonstrate how the wheel spins and stops, with the intention of communicating the causal structures at work. If there is enough common knowledge about the meaning of Dan's gestures, then this becomes an impromptu sign language for conveying explanatory understanding to an audience.

Consequently, sympathizers of tacit understanding owe us some account of how Fred's explanation in a sign language steeped in tradition differs from Dan's allegedly non-explanatory understanding expressed in an improvised sign language. If no principled distinction exists, then Dan has *explained* oscillation using only (or mostly) gestures. Consequently, this is an example of understanding *with* explanation.

But this very quickly moves from a worry about arbitrariness to my third and last worry: being explicit is a pretty easy hurdle to clear. If we start granting Dan and his audience even simple conventions about ostension and basic verbal communication (e.g., "If you do this, then that happens"), then it is hard to see why Fred's explaining in a sign language is much different from Dan's explaining in a language composed of words and gestures. Since these kinds of communicative structures are exceedingly pedestrian, I see no reason that Lipton's examples aren't easy fodder for the Explanatory Objection.

If what I've offered here is a fair gloss of the tacit–explicit distinction and there is no substantive difference between Fred's and Dan's expressive capacities, then tacit explanations are possible. If my proposal conflicts with one's preferred tacit–explicit distinction and there is still no substantive difference between Fred's and Dan's expressive capacities, the challenge remains: why is this an example of understanding without explanation? Hence, there is substantial work to be done before tacit understanding without explanation is a workable concept.

5.6.2. Semantic Knowledge

But suppose, instead, that the key difference between the agents in Lipton's examples of tacit understanding and their explanatory counterparts is captured by the latter's ability to verbally express otherwise identical causal knowledge. Furthermore, suppose that explanations require this verbal acuity. I still think that this is bad news for this class of examples as, quite clearly, explainers thereby know something that tacit knowers do not: namely, that certain verbal expressions refer to certain causal structures. Call this *semantic knowledge*. I shall now argue that if explanations must be explicit, then semantic knowledge distinguishes explanatory understanding from an inferior kind of proto-understanding.

Ex hypothesi, explanations must be communicable or explicit. Hence, tacit understanders have grasped some direct explanatory roles but have not grasped an explanation. Just as the Right Track Objection suggests, this yields proto-understanding. However, scientific knowledge frequently advances when this explanatory information becomes communicable. Hence, on this conception of explanation, although we can achieve proto-understanding without explanation, that understanding improves once we acquire the relevant semantic knowledge.

But is semantic knowledge cognitively beneficial? Lipton certainly does not acknowledge it as such. However, it is unclear what resources he has for rejecting it. After all, Lipton's Assumption identifies understanding with the cognitive benefits provided by explanations; all of Lipton's examples of these benefits are kinds of knowledge, and Lipton also requires explanations to be explicit. The only contentious leap I can see here is that the explicitness of explanation might not entail semantic knowledge, so I'll

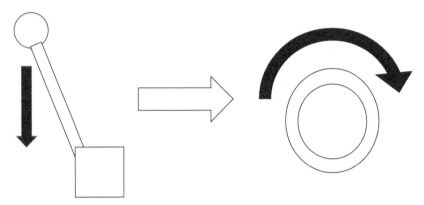

Figure 5.1: Tacit Representation of Causal Knowledge. The white arrow indicates a causal relation.

provide a concrete but simple illustration of Lipton's second example, involving manipulation, to motivate this connection.

Suppose that Dan knows that he can start and stop a wheel by depressing a lever, though he doesn't have words for "wheel" or "lever." Dan needn't even have the word "causes" in his lexicon, but clearly he must have the concept *causes* in order to have causal knowledge. Nevertheless, something like what is represented in Figure 5.1 is within Dan's ken.

Now, contrast Dan with Emily, who can represent this knowledge as Dan does, plus she can express the following in English:

Pressing the lever causes the wheel to spin.

For ease of reference, let's call this "the causal claim." It is hard to see how Emily could be a competent language-user in this context without having the semantic knowledge that *the English sentence, "Pressing the lever causes the wheel to spin," refers to the fact that pressing the lever causes the wheel to spin*. Thus, both she and Dan have propositional knowledge of the causal claim, but Emily also knows that a sentence refers to a causal fact.[13] Hence, if linguistic competence is a mark of explicit knowledge, then the latter implies semantic knowledge.

Let me address two potential objections to my argument. First, one might object that I pay insufficient deference to tacit knowledge, as tacit

---

[13] Here, I follow common conventions: sentences are strings of visual or audible symbols that express propositions; propositions are the abstract entities that carry the meaning of the sentences; facts are concrete things in the world and, unlike sentences or propositions, are not capable of bearing truth or falsity.

knowers often appear to have knowledge that is not easily captured in words. However, the Right Track Objection only requires that at least one explanation can furnish all of the understanding of its counterpart, and these examples are easy enough to construct, e.g.:

> *First Example:* Using an orrery, Andrew tacitly understands why retrograde motion occurs; Belle can explain why retrograde motion occurs by using an orrery.
>
> *Second Example:* By manipulating a complex device, Dan tacitly understands why the device behaves as it does; Emily can explain the device's behavior by manipulating it.

We might think of Belle and Emily as capable of providing "lecture-demonstrations" (i.e., they can do everything that Andrew and Dan can do, and they can provide running commentary on what they're doing). Hence, there is a straightforward recipe for honoring all the precious things that tacit knowers do: we simply imagine them with the ability to say what they know.

However, a final objection remains: one might grant that semantic knowledge is cognitively beneficial, but not in a way that promotes *understanding*. According to the EKS Model, our understanding improves in proportion to its resemblance to scientific knowledge of an explanation. This includes how scientists represent that knowledge. Take a look at a science journal and you'll find many explanations steeped in very precise language. So, from my perspective, this all looks quite rosy.

Of course, those who are moved by the objection should feel no great sympathy with this consequence of the EKS Model. So here are two examples that do not hinge on one's preferred model of understanding. First, throughout the history of science, physical interpretations of technical languages are often required in order to advance scientific understanding, and, when correct, such interpretations are instances of semantic knowledge. Second, in an example that is probably familiar to many readers, consider students who object to their grades on the grounds that their writing does not reflect the depth of their understanding. These complaints frequently ring hollow precisely because verbal expression and understanding cannot be so neatly divorced. In other words, it is natural to think of semantic knowledge as a dimension of understanding.

Thus, we have seen that appeals to tacit understanding face a dilemma. On the one hand, the tacit–explicit distinction is vague enough that the reasons for requiring explanations to be explicit are quite thin. This invites the Explanatory Objection. However, even if that distinction can be forged, then there appears to be some value in making things explicit. Hence, we can follow the advice of the Right Track Objection by treating understanding without explanation as an also-ran and setting our sights on bigger explanatory game that is flush with semantic knowledge.

5.7. Conclusion: A Rapprochement?

In summary, I've used three dialectical strategies to defang Lipton's arguments for understanding without explanation:

- *The Right Track Objection:* In some cases, we can deny that the agents actually understand *why* something is the case. Rather, they possess a modest kind of proto-understanding – they are merely on the right track to understanding why – and this proto-understanding is inferior to explanatory understanding.
- *The Wrong Benefit Objection:* In some cases, we can deny that the understanding invoked in Lipton's arguments is germane to the dialectic.
- *The Explanatory Objection:* In some cases, we can argue that Lipton's examples actually involve understanding *with* explanation.

As we saw, many of these objections stood on different horns of different dilemmas, leaving very little wiggle room for fans of UWE. Hence, ultimately, Lipton's examples pose no threat to the idea that understanding is knowledge of an explanation.

However, I would like to end on a more conciliatory note, by returning to Lipton's Assumption (L1):

> If an explanation of p provides a kind of knowledge about p, then that kind of knowledge amounts to understanding why p.

This assumption raises a puzzle. Given that Lipton countenances understanding without explanation, why does this assumption also accord explanation such a privileged role in understanding?

Let me offer a solution to this puzzle: something in the vicinity of my account of better understanding motivates Lipton's Assumption:

> (EKS1) S_1 understands why p better than S_2 if and only if:
> (A) *Ceteris paribus*, S_1 grasps p's explanatory nexus more completely than S_2; or
> (B) *Ceteris paribus*, S_1's grasp of p's explanatory nexus bears greater resemblance to scientific knowledge than S_2's.

According to the EKS Model, having rich scientific knowledge of a phenomenon's entire explanatory nexus is the *ideal* kind of understanding. Given such an ideal, Lipton's Assumption would be a natural consequence. After all, if you had scientific knowledgeof all of the correct explanations of a given phenomenon, you should presumably also have knowledge of anything (non-trivial) that followed from that knowledge. However, it is precisely this derivative knowledge of an explanation that Lipton's Assumption treats as understanding. Consequently, the EKS Model accounts for the plausibility of Lipton's Assumption.

This is a just-so story at this point. I want to offer two considerations that lend it some further plausibility. First, this would readily account for the pervasiveness of the Right Track Objection presented earlier. Ideals have a way of exceeding the actual. For instance, the fully just society remains forever elusive. Similarly, while even our best current explanations can be improved upon, examples of non-explanatory understanding more starkly highlight the degree to which we fall short of ideal understanding.

Second, if Lipton's Assumption were a consequence of the EKS Model, we would also expect the non-explanatory modes of understanding that animate Lipton's examples to stand in a different relationship to understanding than explanation does in Lipton's Assumption. If explanatory knowledge is the ideal of understanding, then we should expect a certain asymmetry between the cognitive benefits provided by explanatory knowledge versus those provided by Lipton's non-explanatory modes of understanding. Specifically, since an ideal sets a standard, then (trivially) it also always meets that standard. By contrast, everything else meets that standard with varying degrees of success.

For instance, ideal moral agents are moral by definition; the rest of us are, at best, contingently moral. Similarly, Lipton's Assumption entails that knowledge of correct explanations provides understanding by definition. Lipton anticipates this aspect of the EKS Model when he describes his position as one that "lets explanations set the standard for what kind of knowledge counts as understanding" (Lipton 2009, 54).[14]

However, Lipton does not develop another side of this point. Just as morally imperfect people meet a moral standard contingently, non-explanatory practices provide understanding contingently; i.e., it is possible that a person does not understand why *p*, but has:

---

[14] In the same passage, Lipton describes this approach as "narrow." That turns on Chapter 1's earlier observation that procedural understanding is not our concern (nor is it his).

- a merely potential explanation of p,
- a non-explanatory deductive inference that concludes with p,
- an analogy involving p,
- a visual model of p, or
- a manipulation to which p refers.

Consider merely potential explanations. These are propositions that would, if true, explain a phenomenon. By contrast, actual explanations are true potential explanations (i.e., correct explanations). Restricting Lipton's Assumption to *correct* explanations was precisely because some potential explanations, such as conspiracy theories and fairy tales, do not provide understanding.

However, even when potential explanations are not outlandish, they may not provide understanding. For instance, if a doctor misdiagnoses the cause of a patient's symptoms, then, regardless of how reasonable her diagnosis, she misunderstands why the patient has the symptoms she does.

Similarly, while the following deductive inference is sound, it provides no understanding of why parity is conserved in strong interactions:

> Either parity is conserved in strong interactions or unicorns exist.
> Unicorns do not exist.
> ∴ Parity is conserved in strong interactions.

Likewise, many analogies fail to provide understanding, e.g.;

> FOUR: NUMBER:: D: LETTER.

This analogy is perfectly good, but it does not tell us why four is a number, D is a letter, or why they are similar. Similarly, we do not understand why people are happy as a result of knowing that ☺ is a visual model of a happy person.

In many of these cases, *some* kind of understanding may still be in play; e.g., understanding *that* parity is conserved or *that* four is similar to D. But this does not amount to understanding *why* these facts are so. As already stated, only the latter concerns us here. If we recall that procedural understanding is not our concern, parallel considerations apply to manipulations. For example, many people understand *how* to ride bicycles, but far fewer understand the physics that make bicycling possible.

In these examples, merely potential explanations, deduction, analogy, visual models, and manipulation each provide cognitive benefits, but these benefits should *not* be identified with understanding-why. By contrast, Lipton's Assumption expresses the exact opposite when it comes to explanations: their benefits *should* be identified with understanding-why.

Thus, just as the ideal moral agent sets and meets the moral standard by definition, knowledge of a correct explanation achieved through scientific methods sets and meets the standard of understanding by definition. Just as mere mortals are not guaranteed to be moral, so, too, these non-explanatory practices are not guaranteed to provide understanding. Thus, the connection between explanation and understanding is far tighter than the alternative ways of achieving understanding. As a result, the EKS Model provides a plausible rationale as to why there is a special place for explanatory knowledge in understanding even if, as Lipton claims, it is not the only means of achieving understanding.

In conclusion, we might introduce the "Updated Explanation Question:"

• Does understanding improve as one grasps more explanatory information?

Where (relevant examples of) understanding is achieved without explanation, there is always room for explanatory improvement. This is because explanations are cohesive and communicable bundles of direct, critical, and modal information, whereas other forms of understanding seem to be more piecemeal in delivering these goods. Moreover, we have also seen that the central assumption animating Lipton's critiques of explanatory understanding sits comfortably with the EKS Model's claim that *ideal* understanding requires explanation. Hence, even those sympathetic to Lipton's arguments can answer the Updated Explanation Question with a resounding "Yes!" Understanding improves with explanation. Given that few have advocated for the importance of explanation to inquiry more eloquently and ably than Lipton, I hope he would not have disagreed with this conclusion.

CHAPTER 6

Understanding and True Belief

We have now seen that all of the challenges to the received view's privileging of *explanation* – objectual alternatives and myriad examples of putatively non-explanatory understanding – do not pose any significant threat to the Explanation-Knowledge-Science (EKS) EKS Model. Hence, the motivation for a radically novel concept of understanding will have to challenge understanding's status as a species of *knowledge*. From here on out, I show that there is nothing shiny or new to be found here, either: understanding and knowledge are intimately interwoven.

The least controversial aspect of knowledge is that it is *factive*; i.e., that knowledge that p requires p to be true. By contrast, understanding and truth do not always cooperate. The march of science is littered with false theories that nevertheless advanced our understanding. Additionally, science frequently uses highly idealized models to interpret nature's workings, and more than a few have suggested that such idealizations are best regarded as useful fictions.[1] Abject fidelity to detail might well maximize the truths in one's possession, but at the cost of one's understanding – one may lose the forest for the trees. Consequently, some have doubted that understanding tightly tracks the truth (De Regt 2015; Elgin 2004, 2007, 2009a, 2009b; Riggs 2009; Zagzebski 2001). Call these folks *non-factivists* about understanding. Their opponents, *quasi-factivists*, disagree, holding that understanding and truth are more firmly bound (Greco 2013; Grimm 2006; Kvanvig 2003, 2009a; Mizrahi 2012; Pritchard 2007).[2] On their view, idealizations and other departures from the truth are to be explained

---

[1] For example, see essays in Suárez (2009).

[2] Strevens (2008, 2013, 2016) has a nuanced position in this debate. In this chapter, I will mostly use his work without branding him as either a quasi- or non-factivist. Elsewhere, my co-authors and I argue that, while there is some ambiguity in his writings on this front, he is best interpreted as a non-factivist (Doyle et al. manuscript).

away, often as a kind of scaffolding for human cognition that is fragile, fallible, and finite.

Recall that that the received view – championed by philosophers of science from a slightly earlier era – holds that understanding is explanatory knowledge, i.e.:

> *S* understands why *p* if and only if there exists some *q* such that *S* knows that *q explains why p.*

When coupled with the epistemological bromide that knowledge is factive, clear tensions between it and non-factivism arise. These are encapsulated in the Truth Question, first raised in Chapter 1:

* Does understanding require true belief?

Non-factivists answer this question negatively; proponents of the received view and other quasi-factivists answer it affirmatively.

If, like me, you also think that understanding is tethered to knowledge, quasi-factivism appears the more hospitable doctrine. Indeed, my own account of minimal understanding clearly suggests something in the neighborhood of quasi-factivism:

> (EKS2) *S* has minimal understanding of why *p* if and only if, for some *q, S* believes that *q explains why p*, and *q explains why p* is approximately true.

If even *minimal* understanding requires (approximate) truth, then *a fortiori*, so do more demanding instances of understanding. Hence, my Explanation-Knowledge-Science (EKS) Model of understanding appears committed to quasi-factivism. To that end, I consider and rebut two arguments for non-factivism. First, *historical arguments* cite past but false theories that nevertheless enriched our understanding. I argue that historical arguments succumb to the same kinds of objections raised in the previous chapter (Section 6.2). Second, *idealization arguments* appeal to the deliberate misrepresentations and distortions that arise in scientific modeling, but that nevertheless enhance our understanding. I develop an account of acceptance to show how scientists use idealizations without believing falsehoods (Section 6.3).

6.1. Getting Our Bearings

Before proceeding, let me set the stage. What are the main positions in the debate? What are my burdens of proof? As I'll show, the answers to these questions could easily launch us into thornier philosophical thickets than can be reasonably navigated in this book. To avoid those tangles, I will be offering a more programmatic approach, indicating where the thickets arise and how to clear them, without breaking a specific path within this formidable terrain.

6.1.1.   What Is Non-Factivism?

Recall the received view of understanding:

> *S* understands why *p* if and only if *S* knows that *q explains why p*, for some *q*.

Since knowledge is factive, the received view requires understanders to have beliefs in true propositions of the form "*q* explains why *p*." By contrast, non-factivists deny that understanding requires truth in the same way that knowledge does. This amounts to the following:

> *Non-Factivism:* Understanding why *p* does not require belief in any approximately true explanations of *p*.[3]

Quasi-factivism is simply a denial of non-factivism.[4] So construed, non-factivism is a challenge not only to the received view, but also to my EKS Model, given my account of minimal understanding (EKS2, presented earlier).

Three clarifications are in order. First, given my earlier arguments, I will assume that non-factivists assume that the understanding in question is rooted in a false explanation, as opposed to being rooted in something that does not even purport to be explanatory.

Second, this formulation of non-factivism submits to some (too) clever (by half?) counterexamples that I'll be ignoring. For instance, suppose that someone believes in both a true and a false explanation of the same phenomenon but bases his understanding only on the latter. This will contravene the letter of non-factivism but not its spirit. For simplicity's sake, I bracket these sorts of examples in what follows, though small tweaks to the arguments given in Sections 4.2 and 4.3 should handle such cases.

Third, the debate between quasi-factivists and non-factivists concerns the *approximate truth* of statements of the form "*q* explains why *p*." To get a better grip on these "approximately true explanations," recall my "theory" of explanation from Chapter 1:

---

[3] More precisely: $\exists S \exists p(S$ understands why p & $\forall q((S$ believes that *q explains why p*$)\rightarrow$("*q explains why p*" is false)).

[4] More precisely: $\forall S \forall p(S$ understands why $p \rightarrow \exists q((S$ believes that *q explains why p*$)$&("*q explains why p*" is true)).

q (correctly) explains why *p* if and only if:

(1)   *p* is (approximately) true;

(2)   *q* makes a difference to *p*;

(3)   *q* satisfies your ontological requirements (so long as they are reasonable); and

(4)   *q* satisfies the appropriate local constraints.

Recall that the third condition was my way of being as neutral as possible about the scientific realism debate. Despite this neutrality, we can still say something non-trivial about approximately true explanations. An explanation, *q explains why p*, is approximately true if and only if *p* is approximately true, and *some* of the terms in the explanans (*q*) that purport to make a difference to the explanandum (*p*) actually do make a difference and also satisfy your preferred ontological requirements.[5] For a given explanandum, one explanation is closer to the truth (i.e., a closer approximation) than an otherwise-identical rival just in case the former has either more putative difference-makers that satisfy these ontological requirements or fewer putative difference-makers that flout these requirements (*ceteris paribus*). A *strictly* true explanation will be one in which not only *all* of the putative difference-makers satisfy your preferred ontological requirements, but also one that satisfies the other requirements on explanation (true explanandum, satisfaction of local constraints, etc.) without contradiction. I will forfeit a bit of precision for convenience and stipulate that an explanation with no ontologically sanguine difference-making terms is "false."

To illustrate how this works, let's consider two different ontological requirements. A realist-friendly ontological requirement might require a purported difference-maker to refer genuinely. Hence, Ptolemaic explanations will fail this test, but (so far as we know) explanations invoking electrical charge will pass it. By contrast, an antirealist-friendly ontological requirement might be that a putative difference-maker's empirical consequences must be true (and at least some must be non-trivial). As before, Ptolemaic epicycles will fail this test, and (so far as we know) electrical charge will pass it. This is as it should be: any reasonable ontological requirement on our explanantia should accord with good scientific practice. And, of course, we can then plug either of these into the general

---

[5] Without a concrete ontological requirement, this gets a bit murky: do ontological requirements concern explanantia *in toto* or the terms therein? I assume that any ontological requirement designed for one can be revised to accommodate the other.

framework sketched earlier. For the realist, accumulating more difference-makers that genuinely refer brings our explanations closer to the truth. For the antirealist, accumulating more difference-makers that save the phenomena brings our explanations closer to the truth.[6]

Before proceeding, a word of caution: my aims are first and foremost to get clearer about the arguments for non-factivism, recognizing full well that there are dragons at every corner. Turn one corner, and you'll stare headlong into a larger debate about scientific realism; turn another, and you'll be fending off demands for more detailed proposals concerning the semantics of "explains." Instead of slaying those dragons, I'll forge weapons that more valiant philosophers can wield against them. Hence, I will enumerate *possible* objections to non-factivist arguments without a full-throated endorsement of which of these objections I find best. Yet, if this chapter runs as planned, non-factivists will often find more obstacles to resisting the EKS Model than they had before. More interestingly, some non-factivists will find my brand of quasi-factivism so mild-mannered that they can get on board with the EKS Model. Hence, the net result will be a redistricting of the dialectical landscape that makes the EKS Model more formidable to some non-factivists and more inviting to others.

6.2. Historical Arguments

With these clarifications in hand, let me turn to the first sticking point between quasi- and non-factivists: how past theories provide understanding (De Regt 2015; Elgin 2007, 2009b; Kvanvig 2003, 2009a; Mizrahi 2012; Pritchard 2007). Some non-factivists use our benighted heroes from the history of science to motivate a historical argument. For instance, Copernicus's theory falsely describes planetary orbits as circular but still marks a major improvement in our understanding over its Ptolemaic predecessor. Similarly, Kepler and Newton correctly describe these orbits as elliptical but falsely assume that space is absolute and Euclidean. The general structure of such an argument is:

Historical Argument

H1. Some past scientists believed in false explanations of a phenomenon.

H2. These scientists nevertheless had some understanding of why that phenomenon had certain properties.

---

[6] Admittedly, many antirealist-friendly restrictions on explanantia aren't what we typically think of as *ontological* requirements. This is merely terminological.

NF. ∴ *Non-Factivism:* Understanding why *p* does not require belief in any approximately true explanations of (H1, H2).

In what follows, I first flesh out this argument with De Regt's (2015) discussion of the understanding provided by the phlogiston theory (Section 6.2.1). Then, I rebut this argument (Sections 6.2.2–6.2.3).

6.2.1. *Phlogiston*

De Regt provides the clearest version of a historical argument. He argues that factivists are forced to accept the consequence that "we do not have any explanatory understanding at all," for "not only many past theories but plausibly also our current theories are false" (De Regt 2015, 3790).[7] Since De Regt rejects this consequence, he embraces non-factivism.

Of course, De Regt realizes that not any falsehood provides a correct explanation. For instance, simply inferring phenomena from your favorite myth or conspiracy theory and some cleverly gerrymandered auxiliaries does not produce understanding. Hence, he adds that "[e]xplanatory understanding of phenomena requires that theories are *intelligible*, where intelligibility is defined as the value that scientists attribute to the cluster of qualities of a theory that facilitate its use" (2015, 3793). A theory may be intelligible for one scientist and unintelligible to another because only the former has the skills to use it.[8] More importantly, a theory may be intelligible but false:

> The understanding that science provides is rooted in the ability to use and manipulate the model in order to make inferences about the system, to predict and control its behavior [i.e., in intelligibility]. It is in this sense that understanding is a skill rather than a species of knowledge. And it is this characteristic of understanding that allows for the possibility that unrealistic models and false theories can still provide understanding. (De Regt 2015, 3789)

De Regt is willing to bite some sizable bullets in order to push this claim through. The biggest of these bullets is his claim that phlogiston theorists had explanatory understanding of some relevant chemical phenomena.

For the unacquainted, at the dawn of the Chemical Revolution, chemists such as Joseph Priestley posited phlogiston as a fundamental

---

[7] Laudan (1981) provides the *locus classicus* of this so-called pessimistic induction.

[8] De Regt cites his earlier work (2009b) as fleshing out this proposal. There, an explanation must not only be intelligible in the sense just described, but must also meet the "accepted logical and empirical requirements." See Section 2.5.5 for my discussion of this latter requirement.

substance (or "principle") of inflammability. Thus, combustible materials were thought to be rich in phlogiston. Metals were also thought to be rich in phlogiston, for when they are dephlogisticated (when they rust), they lose many of their properties (such as their shininess). However, dephlogisticated metals, or calxes, can recover these properties when they are smelted, which was taken to be a process of (re-) phlogistication. On the conventional historical retellings, the oxygen theory supplanted the phlogiston theory: combustion and rusting did not involve the release of phlogiston, but instead involved the consumption of oxygen. So, we can clearly see how De Regt justifies the first premise of his historical argument: that phlogiston theorists believed in false explanations (H1).

De Regt's remaining claims about understanding rest on a less conventional history: Chang's (2012, ch. 1) careful arguments that the phlogiston theory was abandoned too early during the Chemical Revolution of the late eighteenth century. Chang's work suggests two possible sites of phlogiston-tinged understanding. First, phlogiston theorists had several notable *experimental successes*. For example, Cavendish discovered how to isolate hydrogen by dissolving metals into acids. On his view, the phlogiston in metals was released by the acid. Hydrogen – highly flammable – was taken to be phlogisticated water. Similarly, water was formed by combining phlogisticated water (hydrogen) and dephlogisticated water (oxygen). On this view, phlogiston was "cancelled out" in the process. Finally, Priestley used a burning lens to heat a sample of "mercury calx" in a closed container. On Priestley's view, the mercury calx absorbed the phlogiston, such that the remaining air was dephlogisticated. In our contemporary idiom, Priestley had extracted oxygen from mercury oxide.

These experimental successes indicate that the phlogiston theory was useful and enjoyed non-trivial empirical support. Given De Regt's emphasis on intelligibility's deep tie to manipulation and prediction, we can see why he holds that phlogiston theorists understood why many of these experimental results obtained (H2). Hence, De Regt provides an interesting version of the historical argument.

Chang's work provides a second, and slightly more complex, basis for building a historical argument. By their own standards, which differed substantially from oxygen theorists' standards, phlogiston theorists had many significant *explanatory successes* (Chang 2012, 22–29). The differences in explanatory standards were fourfold. First, when compared to oxygen theorists, phlogiston theorists prized a more complete or comprehensive inventory of explanations, even if those explanations became "cumbersome" or "complex." Second, phlogiston theorists emphasized conservatism, while oxygen theorists prized novelty. Third, while phlogiston and oxygen theorists nominally agreed that unity, systematicity, and empiricism were criteria of explanatory goodness, they differed in their interpretations of these criteria. Fourth, phlogiston theorists adopted a primarily "principlist" approach, according to which chemical substances were to be described and explained in terms of "fundamental substances that impart certain characteristic properties to other substances" (Chang 2012, 38). By contrast, oxygen theorists subscribed to "compositionism," according to which chemical substances were to be described and explained in terms of the elements that constituted them.

According to their preferred constellation of explanatory criteria, phlogiston theorists better explained many of the same phenomena than did the oxygen theory. Additionally, only the phlogiston theory explained why metals shared several common qualitative properties, including "shininess, malleability, ductility, electrical conductivity" (Chang 2012, 21). The oxygen theory failed to explain these properties, though it fared considerably better with quantitative, measurable properties – most notably weight. Indeed, even as the oxygen theory gained prominence in these gravimetric applications, phlogiston theory remained a more fruitful theory in accounting for energy conservation and especially electrical phenomena. Thus, we can see how phlogiston theorists found their theory highly intelligible (in De Regt's sense): it had a cluster of qualities that facilitated its use. So, once again, we have reason to think that a false theory provided understanding (H2).

In De Regt's example, as in other historical arguments, non-factivists are effectively claiming that a merely potential explanation provides actual understanding-why. So construed, this falls squarely into one of the preceding chapter's categories of so-called understanding without explanation (see Section 5.3). As I argued in the previous chapter, these kinds of examples can be neutralized by three possible arguments:

- *The Right Track Objection:* In some cases, we can deny that the agents actually understand *why* something is the case. Rather, they possess a more modest kind of "proto-understanding" that is surpassed by a more demanding kind of explanatory understanding.
- *The Wrong Benefit Objection:* In some cases, we can deny that the understanding invoked in non-factivist arguments is germane to the dialectic.
- *The Explanatory Objection:* In some cases, we can argue that non-factivist examples actually involve understanding *with* explanation.

Furthermore, we saw that these objections conspired to form a network of loosely interlocking dilemmas. To that end, I will apply a similar argumentative strategy to the kinds of cognitive achievements that Chang and De Regt credit to phlogiston theorists.[9] As I'll argue, this allows quasi-factivists either to join non-factivists in biting these bullets or to dine alone on milder delicacies.

6.2.2.   Are the Explanations False?

Those of us who balk at De Regt's conclusion typically deny H2. Consider De Regt's claim that phlogiston theorists had some explanatory understanding of the relevant phenomena. Just as we did in the previous chapter, we can highlight an initial implausibility with this conjecture: it requires the attribution of understanding why p to agents who possess only incorrect answers to the question, "Why p?" For instance, it seems natural to say that people who reply to the question, "Why do metals rust?" or "Why do objects combust?" with the answer, "Because they release phlogiston," *mis*understand why metals rust. So, contra H2, phlogiston theorists don't understand *why* in this case.

While I have strong intuitions that this is the right verdict, perhaps those intuitions are already saturated with quasi-factivist commitments. So, suppose for the sake of argument that we deny that this intuition has any force. Then, while phlogiston theorists' understanding would be preserved (H2), the answers to the aforementioned why-questions would appear to be correct, so that the other premise in the historical argument – that phlogiston theorists only believed in false explanations (H1) – would be in bad shape. In effect, it becomes a version of the Explanatory Objection, in which a putative example of understanding without explanation is challenged on the grounds that it actually is a case of understanding *with* explanation. In the context of non-factivist arguments, this amounts to showing that an explanation alleged to be merely potential is in fact approximately true.

However, there is an "antirealist twist" to this particular version of the Explanatory Objection. On such a proposal, "Metals rust because they release phlogiston" would be (approximately) true. This is possible because non-factivists take the truth-conditions for explanatory statements to be

---

[9] Importantly, this is not a comprehensive discussion of Chang's larger project, only its bearing on De Regt's non-factivism.

less demanding than traditionally thought.[10] To see this, recall my "theory" of explanation presented earlier. All but the third condition are common ground between realists and antirealists; they differ precisely on their preferred ontological requirements for an explanans. So, suppose, following De Regt, you gloss this third condition as follows:

(3*) q is part of an intelligible theory.

Furthermore, suppose that the explanation of p by q satisfies this and the three other conditions. Then antirealists could consistently assert that "q explains why p" is (approximately) true while also asserting that "q" falls short of being approximately true. In other words, because antirealists have more relaxed ontological requirements on the *explanans*, the bar for being an approximately true *explanation* is easier to clear. Such a view would render the historical argument unsound but could easily deliver the same historical judgments as a non-factivist. This immediately accommodates all of the phlogiston theorists' explanatory successes. Hence, some versions of scientific antirealism are consistent with quasi-factivism about understanding.

Having said this, I stress that one needn't be an antirealist in order to brandish the Explanatory Objection; realists can also wield it. For instance, one may argue that the phlogiston explanations traffic in at least some genuinely referring difference-makers. Indeed, while Chang (2012, 246–248) expresses some reservations about explaining the success of science in terms of referential success, many of his remarks are quite congenial to the idea that phlogiston theorists were successfully referring to entities, e.g.:

- " ... the phlogistonist account actually has a close resonance with the modern notion that all metals share metallic properties because they all have a 'sea' of free electrons. If we were to be truly whiggish, we would recognize phlogiston as the precursor of free electrons." (44)
- "Whiggishly speaking, phlogiston served as an expression of chemical potential energy, which the weight-based compositionism of the oxygenist system completely lost sight of." (46)
- " ... how would Lavoisier have done what he did, if Priestley hadn't made oxygen and showed him how to do it, and if Cavendish hadn't made water from hydrogen and oxygen and let Blagden tell Lavoisier about it?" (49)

---

[10] It is an interesting question whether this maneuver should be seen as a revisionary or descriptive semantics for explanatory statements. Answering this question takes us too far afield.

With a little bit of Putnamian polish, realists might regard "phlogiston" as picking out certain properties of free electrons and chemical potential energy. Similarly, "phlogisticated water" and "dephlogisticated air" might well refer to hydrogen and oxygen, respectively.[11] These terms often figured in phlogiston theorists' explanations as difference-makers. As a result, the realist may claim that, insofar as the phlogiston theory explains the common qualitative properties of metals, it is because "phlogiston" refers to free electrons, which in turn make a difference to whether or not metals have these properties. In this case, the Explanatory Objection goes through, thereby undercutting the claim that the scientists in question only believed in false explanations (H1).

It's tempting at this point to ask whether the antirealist- or realist-friendly version of the Explanatory Objection is better. Here, we sail into the tricky dialectical straits that I am trying to avoid: as mentioned earlier, rehashing the long-standing debate between scientific realists and their critics exceeds this book's scope. To that end, I play no favorites about which version of the Explanatory Objection is correct – pick your poison. The crucial point for my purposes is that if the details of either view could be worked out to satisfaction, they would deliver the same judgments about the history of science as the non-factivist, but would be consistent with both the received view and with the EKS Model's account of minimal understanding. For all practical purposes, the tension between non-factivism and my account of understanding would be rendered illusory.

6.2.3. Wrong Benefits and Right Tracks

Recall that the Explanatory Objection hinges on granting the assumption that people who believe that metals rust because they contain phlogiston have genuine understanding of why metals rust. Suppose instead that we assume phlogiston theorists do *not* understand why metals rust or objects combust for precisely this reason. Then H2 is false. More interestingly, we can use the Right Track and Wrong Benefit Objections to explain away the non-factivist intuitions pulling in the opposite direction. Furthermore, unlike the Explanatory Objection, we can do so without tipping our hands with respect to our stance in the scientific realism debate.

First, consider the Wrong Benefit Objection. This objection underscores that the kind of understanding in question is more profitably construed as

---

[11] Admittedly, Chang worries about whiggishness in these passages. It is not obvious to me that realists interested in understanding must adopt the same historiographic scruples.

involving a cognitive benefit other than the explanatory understanding that characterizes the EKS Model's chief target. Phlogiston theorists' experimental successes readily lend themselves to this interpretation. For instance, some phlogiston theorists (e.g., Cavendish) understood *how* to isolate hydrogen from metals. This grants phlogiston theorists a kind of procedural or practical understanding-*how* while denying them the kind of explanatory understanding-*why* that is the proper target of the EKS Model. Such an interpretation accords more naturally with De Regt's emphasis on *skills* as the defining feature of the kind of understanding that is his focus. Zooming out a bit, since understanding how to perform a task and understanding why something is the case are distinct kinds of cognitive achievements, the EKS Model's failure to do justice to the former doesn't impugn its adequacy as an account of the latter. Furthermore, we have roughly and readily explained away non-factivist intuitions that phlogiston theorists understand why metals rust. Such intuitions confuse understanding-why with other sorts of cognitive achievements (in this case, understanding-how).

The remaining strategy in the last chapter was the Right Track Objection. As with the Wrong Benefit Objection, it denies that the false parts of phlogiston theories provide explanatory understanding. However, it diverges from the Wrong Benefit Objection in crediting the agents in question with being on the *right track* to understanding why: they enjoy a kind of *proto*-understanding. As this involves a fair amount of machinery rehearsed in Chapter 4 and briefly summarized in Chapter 5, I will only provide a sketch. Roughly, proto-understanding requires less than a corresponding instance of explanatory understanding. In such cases, it will involve scientifically licit attitudes about different propositions' *explanatory roles*. Explanatory roles are *direct* if they are components of a correct explanation, and they are *indirect* if they play no direct role but instead play a role in scientific knowledge of a correct explanation (e.g., in how scientists consider and compare competing explanations of a given phenomenon).

In the context of the current discussion, the crucial point is that some inquirers may grasp a proposition's explanatory role without grasping the larger explanation in which that role is played. For instance, phlogiston theorists generally had more significant experimental discoveries than oxygen theorists (e.g., discovering the conditions wherein oxygen and hydrogen would be released). The Right Track Objection maintains that while they failed to explain these phenomena correctly, they recognized that the phenomena needed to be explained and that these phenomena would prove useful in adjudicating between competing accounts of combustion. Hence, they identified propositions with important explanatory roles.

In the Wrong Benefit Objection, one enjoys some cognitive achievement *other than* explanatory understanding. By contrast, in the Right Track Objection, one enjoys a cognitive achievement – proto-understanding – that *falls short* of explanatory understanding. Specifically, if one grasps a proposition's explanatory role but cannot situate that proposition within a correct explanation, then one's understanding is inferior to an otherwise-identical person who can weave that proposition into a more comprehensive and correct explanatory story. For this reason, proto-understanding without explanation is always second-rate.

Nevertheless, we can use the Right Track Objection to once again give phlogiston theorists some credit for their achievements, while denying that they have (full-blown) explanatory understanding. Insofar as phlogiston theorists' empirical successes play some explanatory role (e.g., they discovered or recognized certain phenomena that needed to be explained or discovered some genuinely referring difference-makers), we can say, "Phlogiston theorists did not understand why objects combust, but they were on the right track." So, once again, it appears that there are more pedestrian interpretations of the science afforded by a finer-grained taxonomy of understandings that explain away non-factivist intuitions. Moreover, this provides another means of explaining away non-factivist intuitions: they rest on confusing explanatory understanding with something that falls short of the real McCoy.

So, to summarize, either offering significantly inaccurate answers to why-questions affords explanatory understanding or it does not. If it does, then the Explanatory Objection provides ample resources for claiming that this is nevertheless a case of understanding *with* an approximately true explanation. If it does not, then there is no conflict with the EKS Model. Moreover, the Wrong Benefit and Right Track Objections help to explain away any countervailing intuitions. Regardless of which of these three tactics one adopts, historical arguments fail to undergird non-factivism.

6.3. Idealization Arguments

Another argument for non-factivism concerns idealizations' role in understanding. Idealizations misrepresent a system of interest, often for the express purpose of getting a better understanding of that system. Moreover, they are ubiquitous in scientific practice. Some non-factivists rest their case on just these points (Bokulich 2008; De Regt 2015; Diéguez 2013, 2015; Elgin 2004, 2007, 2009a, 2009b; Hindriks 2013; Rohwer and Rice 2013; Zagzebski 2001). By contrast, quasi-factivists find this claim too bold (Greco 2013; Kvanvig 2009a; Mizrahi 2012; Strevens 2008).

To adjudicate between these positions, let's consider a second kind of argument for non-factivism:

Idealization Argument

I1. Some scientists accept idealized explanations of a phenomenon.
I2. All idealized explanations are false.
I3. These scientists nevertheless have some understanding of why that phenomenon had certain properties.
NF. ∴ *Non-Factivism:* Understanding why *p* does not require belief in any approximately true explanations of *p*. (I1–I3)

The most discussed example involves the ideal gas law.[12] To that end, I begin with a brief retelling of this example and then proceed to critique the idealization argument.[13]

We gain understanding of the ideal gas law in terms of the statistical mechanics of the underlying microscopic theory. On this approach, the ideal gas law follows from the partition function. The latter is given by a sum over all states of the system in terms of the energy E of each state:

$$Z = \sum e^{-E/kT}, \qquad (\text{Eq.}\,1)$$

where k is Boltzmann's constant and T is the temperature of the gas.

By assuming that the system consists of N non-interacting particles, each particle's available phase space becomes proportional only to the system volume V. Hence, the partition function will depend on volume as V^N; i.e.,

$$Z = V^N f(T)^N, \qquad (\text{Eq.}\,2)$$

for some function *f(T)*. This result, in turn, leads directly to the ideal gas law:

---

[12] Sullivan and Khalifa (manuscript) review a wider compendium of idealization arguments and find that they all fall short of establishing that understanding is non-factive.
[13] My exposition of this example draws heavily on Doyle et al. (manuscript).

$$PV = VkT\frac{d \ln Z}{dV} = NkT, \qquad \text{(Eq.3)}$$

where P denotes the pressure of the gas.[14]

Assume that the explanation answers the question, "Why does $PV = NkT$?" Then the ideal gas law (Eq. 3) is our explanandum.[15] The explanans consists of the partition function (Eq. 1) and the false assumption (i.e., idealization) that the particles are non-interacting. This idealized explanation is scientifically acceptable (I1) and provides understanding of why the ideal gas law obtains (I3). Moreover, let's provisionally grant that if an explanation has a falsehood in its explanans – as in this case – then the explanation is false (I2). Hence it would appear that this derivation of the ideal gas law fuels an idealization argument.

In principle, the same three strategies from the previous section could be applied here. Idealizations might afford cognitive benefits that play no obvious explanatory roles, in which case the Wrong Benefit Objection applies. Alternatively, they might play an explanatory role but only provide a warmup to proper explanatory understanding. This, of course, would be a version of the Right Track Objection. Finally, aficionados of the Explanatory Objection might argue that idealizations only explain if they are situated within a broader network of more accurate/empirically successful representations – the net result of which is an approximately true explanation. As there are many kinds of idealized explanations, I suspect that these strategies will sometimes be favorable to the ones that I use herein to tackle the ideal gas law.

However, for dialectical purposes, I'd like to set up an especially hospitable scenario for an idealization argument. To that end, I will take I3 to be uncontroversial in everything that follows. This effectively blocks both the Wrong Benefit and Right Track Objections. Furthermore, I'll adopt a decidedly demanding (and realist) notion of approximate truth, so that the antirealist-friendly version of the Explanatory Objection is also not a live option. Specifically, I'll assume, for the sake of argument, that *referential success* is everybody's favorite ontological requirement on explanantia. On this hypothesis, *all* of a *strictly* true explanation's putative difference-makers

---

[14] Note that Nk is equal to nR, where n is the number of moles of gas and R is the ideal gas constant.

[15] Some people prefer to construe the explanandum as the fact that some gases behave in accordance with the ideal gas law (at low density and high temperature). The discussion throughout this chapter would be unaffected by this more cumbersome locution. A frequent mistake is to think that the explanandum is simply the "behavior of gases." As Doyle et al. (manuscript) argue, this coarse-grained construal invites more confusion than insight.

genuinely refer, only *some* of an *approximately* true explanation's putative difference-makers refer, and *none* of a *false* explanation's putative difference-makers refer. We thus get closer to the (explanatory) truth as our explanations cite more genuinely referring difference-makers and eschew any of their non-referring counterparts. If approximately true explanations still emerge unscathed by idealization arguments under these demanding ontological requirements, then we can assume that purveyors of more relaxed approaches to explanatory correctness can coast through these arguments with even lighter fanfare.

With this in mind, I suggest that two safeguards against idealization arguments are still available. First, according to the *Splitting Strategy*, scientists *accept* idealizations but *believe* approximately true explanations, so the term "idealized explanation" illicitly lumps together its two components. Clearly, if workable, such a strategy would render idealization arguments unsound.

However, suppose instead that "idealized explanations" are not the figment of incautious categorizations, and scientists do indeed accept them *en bloc*. In other words, scientists cannot cordon off the idealizations that they accept from the explanations that they believe. Even in this case, which seems more congenial to the idealization argument, my second objection, the *Swelling Strategy*, is a live option. According to this strategy, we expand the notion of scientific knowledge, so that it works just as well with acceptance as with belief. In other words, while my original account of scientific knowledge of an explanation was (roughly) construed as a scientifically warranted *belief* that could not easily have been *false*, the notion of acceptance needed for this version of the idealization argument to work would make it feckless if we expanded the concept of scientific knowledge to include scientifically warranted *acceptance* that could not easily have been *ineffective*. Here, effectiveness is to acceptance as truth is to belief. If the Swelling Strategy works, then the idealization argument would be sound, but a small revision to the EKS Model would readily accommodate it.

Thus, on either strategy, idealizations pose no deep threat to the EKS Model. Clearly, the wellspring of both replies is an account of acceptance, so I begin there (Section 6.3.1). After that, I turn to the Splitting Strategy (Section 6.3.2) and then to the Swelling Strategy (Section 6.3.3).

6.3.1. Acceptance

What does it mean to accept a proposition rather than to believe it? I propose the following account of acceptance:

A person accepts a proposition p in context C if and only if:

> (A1) She adopts a policy of including p among her premises for the purposes of deciding what to do or think in C.

However, to address the idealization argument, we need a more specific account of acceptance that pertains to understanding. In particular, given the EKS Model's privileging of *scientific* knowledge as the source of understanding, I propose the following addendum:

> (A2) The purposes in C are scientific goals.

In the Splitting Strategy, the accepted proposition will be an idealization (e.g., that particles in ideal gases don't interact), while in the Swelling Strategy, the accepted proposition will be an explanation that features an idealization (e.g., that the combination of the partition function and the aforementioned idealization explains why the ideal gas law obtains).

The first condition (A1) draws from Cohen's (1992) canonical account of acceptance. According to Cohen (1992, 4), "to accept that p is to have or adopt a policy of deeming, positing, or postulating that p – i.e. of including that proposition or rule among one's premises for deciding what to do or think in a particular context."

Recall that idealization arguments target my account of *minimal* understanding. Hence, we will want a relatively modest notion of acceptance. Closely related, we will also need to flesh out the success conditions for this modest notion of acceptance. As mentioned earlier, I stipulate that truth is to belief as "effectiveness" is to acceptance. Then Cohen's approach suggests a natural account of effectiveness. Roughly, accepting p is effective if including p as a premise or rule in one's inferential policies plays some non-trivial role in achieving the purposes at hand.

Importantly, just as truth is frequently beyond the ken of many believers, so too is effectiveness beyond the ken of many accepters. Indeed, since I'm staking out a notion of minimal understanding, I will not assume that acceptance of p requires a belief that using p in one's inferential policies is effective. It's simply enough to *be* effective, even if one does not *believe*, much less *know* that this is effective.

In addition to belief being measured by truth and acceptance being measured by effectiveness, these propositional attitudes differ in two other ways. First, while beliefs are also used as premises, they are not sensitive to context-specific purposes. For instance, even though it is effective to accept that particles do not interact in an ideal gas, it is ineffective to accept this idealization when deriving the van der Waals equation, since the van der

Waals equation has specific parameters that represent these interactions. By contrast, it is false to believe that particles fail to interact in ideal gases and it is also false to believe that they don't interact in van der Waals gases. Second, the purpose-sensitivity of acceptance also means that certain consequences of the accepted proposition – namely those that don't matter for the purposes at hand – can be false without sullying the effectiveness of the acceptance. By contrast, if consequences of a believed proposition are false, this clearly sullies the truth of the belief, regardless of those consequences' relevance to the purposes at hand.[16]

These aspects of Cohen's account are fine as far as they go, but they were not intended to bear on the finer details of understanding's relationship to idealization. In particular, even if accepting a claim is effective in advancing some context-specific purposes, those purposes may not be germane to understanding why something is the case. For instance, if advancing a religious mythos about the ideal gas law is the relevant purpose, then there's a good chance that using the idealization concerning non-interacting particles won't be effective and that other idealizations involving deities will be effective. However, neither of these is a desirable consequence. To that end, my addendum to Cohen's account (A2) requires the purposes at hand to be *scientific goals*.

What makes a goal scientific, particularly in the context of explaining? While I will offer no general formulation, here is a small selection of what I have in mind:

- *Answering a particular audience's why-question:* Perhaps the most dis-cussed role of context-specific purposes that pertain to scientific expla-nation come from so-called erotetic theorists (Achinstein 1983; Faye 2007; Garfinkel 1981; Risjord 2000; van Fraassen 1980). The core idea is that the same why-question may admit of different answers in different contexts, depending on the audience's interests, background knowl-edge, inferential capacities, and the like. These sorts of considerations become especially salient when we compare how scientists explain the same thing to (a) specialists in their own field, (b) specialists in other fields, and (c) different kinds of non-specialists.
- *Prediction:* Many, if not all, scientific explanations are proposed with an eye toward having novel testable consequences (Douglas 2009). As we

---

[16] Cohen (1992, 4) defines belief as follows: "belief that *p* is a disposition, when one is attending to issues raised, or items referred to, by the proposition that *p*, normally to feel it true that *p* and false that not-*p*, whether or not one is willing to act, speak, or reason accordingly." I prefer to distinguish belief from acceptance in the ways just glossed.

saw in Chapter 2, alternative explanations often serve to provide crucial experiments that secure (or potentially revise) one's belief in the leading explanation of the time.

- *Control:* Paradigmatically, explanations are causal, and causal information is often useful for the purposes of intervention and manipulation (Woodward 2003). Once again, this is especially important in the quintessentially scientific context of controlled experimentation. However, it is also useful in more applied contexts, such as engineering and biomedicine, where we can expect even more practical interests – such as building a useful device or curing a disease – to give an even richer inventory of scientific goals.

Undoubtedly, there are other scientific goals. Indeed, the next section discusses a fourth example. For the most part, I think that these should be read off of scientific practice. This has benefits and costs. The benefit is that scientific goals can be as diverse (or universal) as good science permits. Looking ahead, this means that the Splitting and Swelling Strategies have the potential to apply to a wide variety of different examples. The cost, of course, is that very little can be said in the abstract about these goals, which makes the proposal a bit hand-wavy. This caveat notwithstanding, the key point is that scientific practice puts an important constraint on whether an idealization is being put to good use and is thereby acceptable.

I conclude my discussion with two passing remarks. First, I am not the first to ally understanding and acceptance. Elgin (2004, 116) states this clearly:

> I do not then claim that it is epistemically acceptable to believe what is false or that it is linguistically acceptable to assert what is false. Rather, I suggest that epistemic acceptance is not restricted to belief.[17]

While our accounts of acceptance are quite similar, Elgin and I put these accounts to different dialectical purposes. First, my account of acceptance is specifically tailored to the interplay of idealization and explanation. By contrast, Elgin takes acceptance to apply to a much larger stable of cognitive practices. Furthermore, and more strikingly, Elgin's account of acceptance undergirds her non-factivism; mine, my (soft-spoken) quasi-factivism.

---

[17] Elgin has been misinterpreted on this point. For instance, Pritchard (2007) takes Elgin to assert that in the case of idealizations, "scientists ... have false *beliefs* in the subject matter." He then goes on to suggest that "scientists might well *accept* their theories in such cases (i.e. endorse them for all practical purposes, as the best theory available), even though they don't actually believe them." Curiously, this comes three years after Elgin made precisely the same distinction.

Second, recall that Chapter 1 mentioned several dimensions of resemblance to scientific knowledge that improve our understanding; e.g., the number of plausible potential explanations that the person has considered, the number of considered explanations that the person has compared using scientifically acceptable methods and evidence, the scientific status of the methods and evidence that the person used to compare the explanations, the safety of the person's beliefs about explanations, and the accuracy of the explanations that the person believes. My discussion of acceptance suggests that resemblance to scientific knowledge is also proportional to the following:

- The variety of ways that the person can use explanatory information (broadly construed to include the relevant idealizations) so as to achieve different scientific goals

The thought that understanding improves in proportion to the variety of scientifically useful ways that a person can represent an explanation has independent plausibility. Suppose that one person can only derive the ideal gas law without using the idealization, whereas a second can derive the ideal gas law either using the idealization or not. *Ceteris paribus*, the latter's understanding of the ideal gas law outstrips the former's. Hence, acceptance is a useful addendum to the EKS Model.

6.3.2.   *Splitting*

With a clearer notion of acceptance in hand, let's consider the first of our rebuttals to idealization arguments, the Splitting Strategy. According to this strategy, we can distinguish the idealizations that scientists accept from the approximately true explanations that they believe. However, this contravenes non-factivism, for scientists thereby believe in approximately true explanations.

This strategy readily applies to the ideal gas example. Begin with the claim that scientists accept the idealization that particles in ideal gases do not interact. Quite clearly, the assumption about non-interacting particles is being used as a premise in the derivation of the ideal gas law (A1). But which scientific goal does this advance? Here, I will steal a page from Strevens (2008). In the derivation of the ideal gas law, only the partition function, Z, contains the explanatorily relevant *difference-makers*; namely, the energy (E) and temperature (T) of the system. These appear to be genuinely referring terms and hence satisfy the (decidedly realist) ontological requirements which I have adopted for the sake of argument.

Hence, the explanation is approximately true. By contrast, the idealiza-tion – the assumption of non-interacting particles – serves a different scientific goal: it highlights the explanatory irrelevance of otherwise-plausible potential explanatory factors (A2). As a useful term of art, let's say that while explanations cite difference-makers, idealizations flag *difference-fakers*.

To appreciate particle interactions' status as difference-fakers, consider a "de-idealized" explanation of the ideal gas law. Instead of assuming non-interacting particles, such a derivation has parameters that represent inter-molecular interactions. As a result, something far more accurate than the ideal gas law can be derived, namely the virial equation of state:

$$PV/NkT = 1 + B/V + C/V^2 + D/V^3 + \ldots \qquad (\text{Eq.4})$$

This expansion can be derived directly from statistical mechanics and rendered arbitrarily precise by extending the equation indefinitely, with each added term being derivable from increasingly detailed and accurate assumptions about the particle interactions. For instance, B corresponds to interactions between pairs of particles; C, triplets; D, quartets; and so on. From here, we can derive the ideal gas law. In particular, the ideal gas law only holds at relatively low densities (P) and high temperatures (T). Consequently, volume (V) will be large so the contribution of the added terms (B/V, C/V^2, D/V^3, etc.) will be vanishingly small. In other words, the de-idealized explanation shows us *why* particle interactions won't make a difference. However, idealizations are a more effective way of conveying *that* particle interactions don't make a difference. By showing that we can derive the ideal gas law even when we assume that particles don't interact at all, we flag them as explanatorily irrelevant (i.e., as difference-fakers).

There are two crucial lessons to be drawn from the Splitting Strategy. First, recall that I'm aiming to defuse the following premises in the idealization argument:

> I1. Some scientists accept idealized explanations of a phenomenon.
> I2. All idealized explanations are false.

If the Splitting Strategy is right, then I1 should be replaced with the following:

> I1*. Sometimes, scientists understand by both having approximately true beliefs in explanations and accepting idealizations used to represent those explanations.

The preceding is consistent with this revision. In particular, scientists can believe that only the partition function (Eq. 1) explains the ideal gas law, while accepting the idealization that particles in ideal gases do not interact for the purposes of flagging difference-fakers. Since non-factivists have given us no reason to think that the partition function is a false description, the idealization argument fails to establish non-factivism.

Similarly, I2 overstates the case. At best, the preceding only commits us to:

I2*. All idealizations are false.

However, the fact that all idealizations are false does not justify non-factivism, since the needed connection between falsehood and *explanation* has been severed.

Finally, recall also that idealization arguments were tough precisely because I was striving to be maximally charitable to the non-factivist. However, anyone with more modest ontological requirements will have an easier time establishing the approximate truth of "idealized explanations." Furthermore, the framework that I've just presented allows idealizations to play an inexhaustible number of roles in promoting scientific goals – they needn't only flag difference-fakers. Hence, I take these considerations to shift the burden of proof squarely back on non-factivists' shoulders.

6.3.3. Swelling

However, perhaps the Splitting Strategy only works because of the idiosyncrasies of the ideal gas law. Perhaps other idealizations are so deeply entangled in an explanation that the Splitting Strategy is unworkable. While I'm more inclined to think that this is the result of incautious formulation, let me grant that such scenarios are possible, if only for the sake of argument. Even in such cases, the second, Swelling Strategy readily applies. On this line of argument, we simply expand the notion of scientific knowledge so that acceptance and belief are interchangeable components.

Let's begin by clarifying what it would mean to accept an idealized explanation. For expediency's sake, suppose that the idealizations cannot be sequestered from the difference-makers in our explanation of the ideal gas law. Then the idealized explanation is:

$PV = NkT$ because both $Z = \sum e^{-E/kT}$ and the system consists of N non-interacting particles.

Already, there are problems with even motivating a challenge to the EKS Model. This idealized explanation includes the partition function. As just argued, that function cites difference-makers that have genuine referents, namely energy (E) and temperature (T). So, even this idealized explanation is approximately true. As a result, we can deny I2.

However, this leaves two loose threads. First, we can see non-factivism as consisting of two theses: the first denies that understanding requires (approximate) *truth;* the second denies that understanding requires *belief.* Hence, even if non-factivists grant that idealized explanations are approximately true, they can still challenge both the received view and the EKS Model with respect to this second thesis. On this view, non-factivism claims that understanding can be achieved by acceptance in the absence of belief. So, the first loose thread concerns the relationship of understanding and belief, independently of its relationship to truth.

The second loose thread is the real possibility that some idealized explanations are not even approximately true. Perhaps idealizations play such a central role in these explanations that they eclipse the referring difference-makers. On this view, the idealized explanation is, on balance, more false than true, and thereby rob undeserving of its "approximately true" designation.

The Swelling Strategy ties these loose threads together. On this view, acceptance can replace belief, and we should revise the account of minimal understanding accordingly, viz.:

(EKS2*) S has minimal understanding of why p if and only if, for some q,
(A) S believes that q *explains why p*, and q *explains why p* is approximately true; **or**
(B) S **accepts** that q *explains why p*, and q *explains why p* is **effective**.

In this way, the notion of minimal understanding "swells up" to include an acceptance clause that can even host the idealized explanations that take the longest holidays from the truth.

Now that we're clear about what the Swelling Strategy involves, let me admit that it makes me a bit anxious, for if it is the only viable rebuttal to non-factivism, then perhaps philosophizing about understanding forces us to recruit a theory of acceptance that is not a typical staple of the epistemology of scientific explanation. This, of course, runs afoul of the larger objective of the book, which is to render understanding into a relatively unremarkable object of philosophical theorizing.

Having said this, I'll talk myself off the ledge. First, to show that the Swelling Strategy is the only viable rebuttal, we would need to show that the other four – the Explanatory, Wrong Benefit, and Right Track Objections plus the Splitting Strategy – don't work. That's no small task. So, I can take some solace in treating the Swelling Strategy as my nuclear option.

Second, even if the Swelling Strategy is the only viable rebuttal to non-factivism, then its elevation of acceptance need not be regarded as a radical break from the epistemology of *scientific* explanation. As such, the general tenor of the book remains unfazed. This proposal gains further plausibility when we remember that my account of minimal understanding (EKS2) was only one half of the EKS Model. The EKS Model's other half is my account of better understanding:

(EKS1) S_1 understands why p better than S_2 if and only if:
  (A) *Ceteris paribus*, S_1 grasps p's explanatory nexus more completely than S_2; or
  (B) *Ceteris paribus*, S_1's grasp of p's explanatory nexus bears greater resemblance to scientific knowledge than S_2's.

As I have throughout, I will call (A) the "Nexus Principle" and (B) the "Scientific Knowledge Principle." Let's focus on the second of these principles. EKS2* would be a principled revision if acceptance were part and parcel to higher grades of understanding and, in particular, if it were at home in scientific knowledge. We could then see EKS2* as a natural endpoint on a continuum of epistemic statuses that already tolerate acceptance.

With that in mind, note that the Scientific Knowledge Principle provides an intuitive rationale as to why accepting idealizations provides understanding. The principle holds that understanding improves in proportion to its resemblance to scientific knowledge. Idealizations are one of the most distinctive aspects of scientific explanations. Since scientists frequently know that idealizations are false, they do not believe in idealizations but nevertheless use them in their inferential policies. Hence, resemblance to scientific knowledge often involves using idealizations as a scientist would (i.e., accepting them).

Of course, this assumes that acceptance can figure in scientific knowledge. I shall now argue that acceptance can play the same role in scientific knowledge of an explanation as belief. To get a better sense of this, let's remind ourselves how I define scientific knowledge of an explanation:

> *S* has scientific knowledge that *q explains why p* if and only if the safety of *S*'s belief that *q explains why p* is because of her scientific explanatory evaluation (SEEing).

The Swelling Strategy suggests the following revision:

> *S* has scientific knowledge that *q explains why p* if and only if the safety of *S*'s belief **or acceptance** that *q explains why p* is because of her SEEing.

To defend this expanded notion of scientific knowledge, we need to defend the following claims:

- Like belief, acceptance can be safe.
- SEEing can result in either belief or acceptance.

A safe belief is one that could not easily have been false. Several epistemologists take this to be a precise specification of the "anti-Gettier" condition on knowledge (Pritchard 2005; Sainsbury 1997; Sosa 1999; Williamson 2000).[18] By analogy, safe acceptance occurs when one's acceptance could not easily have been ineffective. In other words, if someone's use of a claim results in the fulfillment of particular scientific goals, it is fair to ask whether the manner in which she used this claim is one that is liable to advance those goals or if the person simply got lucky. For instance, consider a slight variation on our idealized explanation:

> The gas in the container behaves as an ideal gas because both $Z = \sum e^{-E/kT}$ and the system consists of N non-interacting particles.

Suppose further that the explanation is intended to advance the complex scientific goal of identifying difference-makers and flagging difference-fakers. Suppose that the person only looks at an ideal gas. Still, if she could have easily looked at containers filled with non-ideal gases, her acceptance will be ineffective: she could have easily flagged particle interactions as difference-fakers when they were *bona fide* difference-makers. Hence, it appears that there is a coherent sense in which acceptance can be safe.

Moreover, acceptance and belief play nearly identical roles in the process of scientific explanatory evaluation (SEEing) that I've discussed in earlier chapters. Recall that SEEing has three invariant features. First, scientists *consider* many of the plausible potential explanations of the phenomenon of interest. Second, using the best methods and evidence, scientists *compare*

---

[18] I discuss safety's role in understanding in Chapter 7.

the potential explanations that have been considered. In paradigmatic cases of comparison, one explanation is the "winner" of these comparisons, though sometimes multiple explanations are good along different dimensions, and often these explanations are complementary rather than competitors. In the last stage of SEEing, scientists *form attitudes* based on the comparisons just discussed. Until now, I have been assuming that scientists *believe* that clear winners in the prior stage of comparison are correct, believe that clear losers are incorrect, and assign appropriate degrees of belief about the middle of the pack.

According to the Swelling Strategy, the formation stage should now include not only these doxastic attitudes, but also non-doxastic attitudes, such as acceptance. We can get a clearer sense of how this works by thinking about how scientific goals enrich the previous stage of comparison. In addition to the question of which of the candidate explanations accords with the evidence, there is a further question: which explanations (or ways of representing an explanation) are most effective in advancing the relevant scientific goals? In the best cases, evidential considerations and scientific goals will accord with each other, and the inquirer both accepts and believes the right explanation. However, acceptance will eclipse belief when certain tradeoffs arise. For instance, a slightly less accurate explanation will sometimes be more effective in advancing communicative, medical, or engineering purposes (e.g., when one's purposes include simpler derivations or the flagging of difference-fakers).

At this point, the only question is whether an agent who safely accepts an idealized explanation as a result of SEEing should be credited with scientific knowledge. I offer three considerations in favor of this view. First, acceptance's nearly identical role to belief's in safe SEEing is already *prima facie* grounds that they are very similar kinds of mental states and hence often serve similar functions in more complex mental states, such as scientific knowledge.

Second, I suspect that, from an ordinary language perspective, tokens of "scientific knowledge" and its cognates frequently accord with this acceptance-based reading. Indeed, it's been noted more than a few times that the standard epistemological account of knowledge as requiring true belief does not accord well with scientific practice. To choose but one example, Kvanvig (2003, 201) writes, "[W]e honorifically talk about the present state of scientific knowledge (even though we know that some of what falls under that rubric is false)." Rather than treating this as loose or "honorific" talk, the Swelling Strategy suggests that we expand our notion of scientific knowledge to accommodate these uses. As we've seen, the expansion appears relatively modest.[19]

Third, acceptance might well be the most common outcome of scientific inquiry, yet I suspect that these outcomes are frequently dubbed "scientific knowledge" nonetheless. Indeed, many of the scientific case studies discussed in this book occur at a relatively early stage in a theory's development, where this is the proper attitude, as scientists wait for more evidence to come in. So, the "honorific-talk maneuver" that Kvanvig and other epistemologists adopt ends up looking increasingly stipulative and *ad hoc* as we survey some of our most significant scientific discoveries.

If, after all of this, epistemologists still think that I'm doing violence to their pet concept, I offer two replies. First, talk yourself off the ledge: the extent to which idealization arguments are defused by combination of the Splitting Strategy, the Explanatory Objection, the Wrong Benefit Objection, and the Right Track Objection has not been determined. Hence, perhaps the Swelling Strategy is otiose, and any hand-wringing would be for naught. Second, even if we need the Swelling Strategy, perhaps the surface structure of our epistemic terms is misleading us, and "knowledge," as used by epistemologists, and "scientific knowledge" do not stand in a genus–species relationship. Instead, perhaps folk and scientific knowledge are either species of a more generic concept of knowledge or only bear family resemblance to each other.[20]

Taking stock, we have seen that, even on a fairly generous reading of idealization arguments, there is no shortage of quasi-factivist blockades. In particular, acceptance can play a perfectly healthy role in a knowledge-based account of understanding. According to the Splitting Strategy, the idealization argument rests on a failure to distinguish scientists' attitudes toward their explanations from their attitudes toward their idealizations. With greater caution, the problem disappears. On the Swelling Strategy, our concept of scientific knowledge expands to include not only safe and scientific beliefs, but also safe and scientific acceptances. Let me reiterate that I have granted things that other quasi-factivists might well resist. Perhaps idealized explanations don't provide understanding-why or aren't nearly as far from the truth as non-factivists advertise. In such cases, the objections rehearsed in the last chapter and in the discussion of the historical arguments come back with a vengeance. As I stated earlier,

---

[19] Moreover, I am far from the first person to suggest that scientific knowledge frequently traffics in acceptance rather than belief (Cohen 1992; Steel 2013; Wray 2001).

[20] Consistent with this thesis is psychological evidence suggesting that scientists have greater tolerance for false explanations than laypeople do (Braverman et al. 2012).

the sensitivity of non-factivist arguments to the examples chosen will always provide a license for optimism. As a result, I suspect that my arguments lack an air of conclusiveness. I consider it a victory if I've filled the toolbox of responses that would put the EKS Model in good standing, so that non-factivists have a clearer sense of the argumentative devices they must disassemble to unseat my view.

6.4. Conclusion

To summarize, I have argued that the leading reasons to be a non-factivist about understanding – concerning the history of science and the use of idealizations – both miss their mark. In some cases, I have highlighted certain weaknesses in non-factivists' arguments. In other cases, I have shown that the conflict between non-factivism and the EKS Model is a will-o-wisp.

As mentioned earlier, this chapter aims to be more suggestive than conclusive – to offer a framework for thinking about non-factive understanding that gestures to how a more comprehensive survey of scientific examples might be wedded to a broader set of dialectical maneuvers than has been previously considered. To that end, it's worth noting that while I've mentioned how the preceding chapter's strategies apply to both historical and idealization arguments, the acceptance that I've associated with idealization arguments may also apply to historical arguments. For instance, it may be that our historical predicament means that scientists ought to accept (rather than believe) many of their best explanations – even those lacking idealizations. Seen in this light, all of the strategies and objections in this chapter are in principle applicable to any putative case of non-factive understanding. As a result, non-factivists of all stripes should consider the various objections presented here as a kind of checklist of obstacles to be overcome: Explanatory Objections, Wrong Benefit Objections, Right Track Objections, Splitting Strategies, and Swelling Strategies.

My argument, in the end, amounts to murder by numbers. There's a healthy chance that at least *one* of these responses to a non-factivist argument will stick. Insofar as this list might be whittled down, it will require greater clarification of non-factivists and quasi-factivists' shared assumptions. If some of these moves encroach on this common ground, then non-factivists' checklists could be abbreviated accordingly. Perhaps this would open further space for non-factivists to operate. However, until then, the balance of arguments favors quasi-factivism. Hence, much like knowledge, understanding requires true beliefs – or, perhaps, their nearby cousins, effective acceptances.

As we have seen, even my nuclear option – the Swelling Strategy – does not require a deep departure from the idea that understanding is an artifact of inquirers' knowing how to put their whys and wherefores to good scientific use. Rather, the career of acceptance in science has been solid, if not storied. Hence, understanding's relationship to truth does not suggest a radical rethinking of its relationship to science, knowledge, and explanation.

CHAPTER 7

Lucky Understanding

It's widely accepted that knowledge entails non-accidental true belief. As I've shown, we should not be afraid to require true beliefs of understanding, so long as we handle the contents of those beliefs with care. However, we still must ascertain whether the near-ubiquitous epistemological conviction that knowledge is "non-accidental," "Gettier-proof," or "luck resistant" holds court in the realm of understanding.

But what kind of luck must knowledge resist? Roughly, the anti-luck intuition prohibits a knower's true belief from being the result of some fluky causal chain. But is it also the case that understanders must have true beliefs that are not the result of such fortuitous arrangements? Here, we reach the last of our questions concerning the nature of understanding: the Classic Luck Question:

- Is understanding incompatible with luck?

Since knowledge is incompatible with luck, proponents of the received view would appear forced to answer this in the affirmative, for as you'll recall the received view asserts:

> S understands why p if and only if there exists some q such that S knows that q explains why p.

While several philosophers follow the received view in answering the Classic Luck Question affirmatively (DePaul and Grimm 2007; Greco 2013; Grimm 2006; Kelp 2014; Khalifa 2011, 2013; Khalifa and Gadomski 2013; Riaz 2015), others disagree (Hills 2015; Kvanvig 2003, 2009b; Morris 2012; Pritchard 2008, 2009a, 2010, 2014; Rohwer 2014; Zagzebski 2001). As a result, they deny that understanding is a species of knowledge.

I will argue that while the Classic Luck Question admits of a negative answer, that answer is harmless to my Explanation-Knowledge-Science

(EKS) Model of understanding (Section 7.2). Of course, this is because I countenance *degrees* of understanding:

(EKS1) S_1 understands why *p* better than S_2 if and only if:

(A) *Ceteris paribus,* S_1 grasps *p*'s explanatory nexus more completely than S_2; or

(B) *Ceteris paribus,* S_1's grasp of *p*'s explanatory nexus bears greater resemblance to scientific knowledge than S_2's.

As such, some understanding is lucky, and some is not. However, it turns out that while luck does not eradicate understanding, it diminishes it. So, the fact that some relatively undemanding instances of understanding can be lucky should not skew our judgments about the whole spectrum of understanding.

This brings us to the more interesting question about luck and understanding, what Chapter 1 called the Updated Luck Question:

- Does understanding improve as it becomes less lucky?

I will argue that when we attend to the details of scientific practice, negative answers to the Updated Luck Question pose no deeper threat to the EKS Model than negative answers to the Classic Luck Question (Sections 7.3–7.5). But, first, let me begin by clarifying precisely what these questions are asking (Section 7.1).

7.1. The Luck Questions

As mentioned earlier, the Classic Luck Question's chief target is the received view:

S understands why *p* if and only if, for some *q*, *S* knows that *q explains why p.*

To make any headway on how luck bears on understanding, we need a relatively clear account of how luck figures in knowledge. To that end, I use a safety-based analysis of luck (Pritchard 2005; Sainsbury 1997; Sosa 1999; Williamson 2000). More precisely, I follow Pritchard's (2009b, 34) Safety Principle:

S's belief is safe *iff* in most near-by possible worlds in which *S* continues to form her belief about the target proposition in the same way as in the actual world, and in all very close near-by possible worlds in which *S* continues to form her belief about the target proposition in the same way as the actual world, her belief continues to be true.

Hence, I will be assuming that a belief is safe if and only if it is not epistemically lucky; i.e., it is not accidental in a way that disqualifies it as knowledge. Note that this claim is compatible with a safe belief failing to be knowledge for other reasons (e.g., by violating some justification or ability requirement).

To be sure, there are other accounts of epistemic luck.[1] I have chosen safety for three reasons. First, Pritchard is one of the foremost proponents of anti-luck epistemology and of lucky understanding. Thus, using his account of epistemic luck has certain dialectical advantages. Second, among the leading accounts of what it means for a belief to be non-accidental, safety is the most modest. For instance, it requires less than Nozick's (1981) sensitivity condition and any other accounts that add further (e.g., ability) conditions to the modal conditions that safety already requires (e.g., Greco 2009; Sosa 1999). Third, safety accords nicely with scientific practice, as I'll argue later.

With this, we have a clear sense of what an affirmative answer to the Classic Luck Question amounts to; namely, the following consequence of the received view:

> *Classic Luck Incompatibilism:* Understanding why *p* requires a safe belief in some explanation of *p*.

By contrast, critics of the received view endorse the following:

> *Classic Luck Compatibilism:* Understanding why *p* does not require a safe belief in any explanation of *p*.

Let's now turn to the issue of lucky understanding in its updated form. The big move, as we've seen, is to think of understanding in terms of degrees, rather than as an all-or-nothing affair. Hence, our debate should shift to evaluating the claim that understanding improves with safety, i.e.:

> *Updated Luck Incompatibilism: Ceteris paribus*, if two people both have a true belief in the same explanation of *p*, the person with the safer belief has a better understanding of why *p*.

Thus, those who deny this view endorse the following:

> *Updated Luck Compatibilism:* There are cases in which two otherwise-identical people both have a true belief in the same explanation of *p*, but the person with the safer belief does not have a better understanding of why *p*.

---

[1] See Orozco (2011) for a broader review of the literature.

As we'll see, both of these positions must be further updated if they are to make direct contact with the EKS Model. This caveat notwithstanding, I'll first argue that updated luck incompatibilism makes quick work of classic luck compatibilism and then argue that updated luck incompatibilism is an altogether plausible doctrine – at least if it hitches its wagon to the EKS Model's star.

7.2.   Passé Compatibilism

Let's begin with the Classic Luck Question. There are two kinds of luck that understanding (but not knowledge) is alleged to accommodate. Epistemologists frequently refer to all examples where luck undermines a putative instance of knowledge as "Gettier cases." By contrast, I will follow Pritchard – a central advocate for lucky understanding – in distinguishing *Gettier-style* luck from *environmental* luck. According to Pritchard's (2009a, 21) dichotomy, in the former cases, bad luck "intervenes 'betwixt belief and fact',"[2] e.g.:

Sheep Example

Abel is a shepherd who looks upon his pasture one day and sees something that looks like a sheep. He thus comes to believe that a sheep is in the pasture. As it turns out, Abel is looking at a shaggy dog that is perfectly occluding a sheep that is actually in the pasture. Had there been no sheep behind the dog, Abel would have falsely believed that a sheep was in the pasture.[3]

Abel's true belief is clearly lucky. To paraphrase Pritchard, the shaggy dog intervenes between the belief and the fact that the sheep is in the pasture. As such, beliefs that are susceptible to this kind of luck are not candidates for knowledge. I relegate discussion of Gettier-lucky understanding to the Excursus (Section 7.6).

For the bulk of this chapter, I focus on *environmental* luck, which provides the foremost luck-related challenge to the idea that understanding is a species of knowledge. Unlike Gettier-style luck, environmental luck does not intervene between fact and belief. Goldman (1976), by way of Ginet, provides its classic incarnation:

Barn Façade Example

Bonnie is driving through a rural county, when she sees a barn-shaped object. Thinking this unexceptional, Bonnie forms the true belief that there is a barn in the field. However, unbeknownst to her, she is driving in

---

[2] Pritchard credits Unger with coining this phrase.   [3] Paraphrased from Chisholm (1966).

Barn Façade County, where most objects that look like barns are in fact fakes. Moreover, had Bonnie looked at a fake barn, she would have believed it was a real barn.

Bonnie's true belief is clearly lucky. However, unlike Abel's shaggy dog, the piece of bad luck – that she is in Barn Façade County – does not intervene between the belief and the fact that a barn is in the field. Nevertheless, it is widely held that environmentally lucky beliefs of this sort do not amount to knowledge.

In what follows, I first present the leading argument for environmentally lucky understanding (Section 7.2.1). I then highlight two defects with that argument (Sections 7.2.2 and 7.2.3).

7.2.1. Pritchard on Environmental Luck

Using analogies with the Barn Façade Example, some luck incompatibilists have argued that, unlike knowledge, understanding is compatible with environmental luck:[4]

Nero Example

Nero comes home to find his house in flames. When he asks a firefighter what caused the fire, she gives him the correct answer that it was a faulty breaker box. Unbeknownst to Nero, the person he asked is one of the few real firefighters on the scene, as many nearby people are dressed as firefighters en route to a costume party. Nero could have very easily asked these partygoers, and, had he done so, they would have given him a false answer while failing to indicate that they were not real firefighters.[5]

Just as Bonnie could have very easily looked upon a barn façade and formed the false belief that a barn is in the field, Nero could have very easily consulted a partygoer and formed the false belief that, e.g., an arsonist caused the fire. Hence, Pritchard reasons that Nero's belief is environmentally lucky. Pritchard (2010, 79) then argues that:

> ... the agent concerned has all the true beliefs required for understanding why his house burned down, and also acquired this understanding in the right fashion. It is thus hard to see why the mere presence of environmental epistemic luck should deprive the agent of understanding.

---

[4] The most prominent arguments to this effect come from (Kvanvig 2003; Pritchard 2010). Hereafter, I focus on Pritchard's argument since it provides the most careful analysis of the relationship between understanding and luck. My treatment of Pritchard readily applies to Kvanvig and, indeed, the proponents of Gettier-lucky understanding. I address the latter in the Excursus.

[5] Paraphrase from Pritchard (2010, 79).

Let's render Pritchard's reasoning explicit:

Environmentally Lucky Understanding Argument

ELU1. Nero understands why the house burned down.

ELU2. If two beliefs are similar in all epistemically relevant respects, and if one of those beliefs is unsafe, then so is the other.

ELU3. Bonnie's belief in the Barn Façade Example and Nero's belief in Pritchard's Nero Example are similar in all epistemically relevant respects.

ELU4. Bonnie's belief in the Barn Façade Example is environmentally lucky.

ELU5. If a belief is environmentally lucky, then it is unsafe.

ELU6. ∴ *Classic Luck Compatibilism:* Understanding why p does not require a safe belief in any explanation of p. (ELU1–ELU5)

Thus, the claim is that understanding, unlike knowledge, tolerates environmental luck. If sound, Pritchard's argument clearly refutes classic luck incompatibilism. By implication, it also unseats the received view.

As mentioned earlier, *updated* luck *incompatibilism* defuses much of what is compelling about *classic* luck *compatibilism*. More precisely, by switching from the received view to the EKS Model, we can undercut ELU1's role in Pritchard's argument (Section 7.2.2). Additionally, I will argue that Pritchard's Nero Example differs in epistemically relevant ways from the Barn Façade Example, thereby challenging ELU3 (Section 7.2.3). As we'll see, both of these points set the stage for my newfangled luck incompatibilism in Section 7.3.

7.2.2.   Degrees of Understanding

Both Pritchard's first premise (ELU1) and conclusion (ELU6) indicate that he is taking understanding in either its outright or its generic form. Recall our analysis of these concepts:

> *Outright Understanding:* "S understands why p" is true in context C if and only if S has minimal understanding and S approximates ideal understanding of why p closely enough in C.

> *Generic Understanding:* S has some understanding of why p if and only if "S understands why p" is true in some context C.

As we've seen throughout, this might score some quick points against the received view, but it's toothless when pitched against the EKS

Model, primarily because the latter accounts for degrees of understanding. When degrees of understanding are taken on board, ELU1 becomes:

> ELU1*. There is some context in which "Nero understands why the house burned down" is true.

It seems quite uncontroversial that there is at least *some* context in which Nero grasps enough explanatory information in a sufficiently "scientific" way. In particular, Nero has, at the very least, *minimal understanding*: an approximately true belief in a correct explanation (see Chapter 3). Note that other contexts will also tolerate luck, but require more than minimal understanding and less than knowledge. Hence, Nero understands in contexts with these low standards.

With the move from ELU1 to ELU1*, Pritchard no longer gets his desired conclusion – that understanding is lucky. Rather, the argument is only valid if we revise the conclusion as follows:

> ELU6*. There are some contexts in which understanding why p does not require a safe belief in any explanation of p.

Whereas Pritchard's original argument gave us classic luck compatibilism, ELU6* highlights where the Classic Luck Question goes wrong: contexts that tolerate unsafe understanding are entirely consistent with understanding *improving* as it more closely approximates knowledge and becomes safer in more demanding contexts. Of course, these points will require further arguments (see Sections 7.3 and 7.4), but, for now, I simply want to point out that some failures of the received view's treatment of luck (*classic* luck incompatibilism) are no slight against the EKS Model (*updated* luck incompatibilism).

7.2.3. *Barn Façades and Firefighters*

In and of itself, the argument against ELU1 already signals the need to update our questions about understanding and luck. Nevertheless, it will be instructive to highlight another difficulty with the Environmentally Lucky Understanding Argument, namely its third premise (ELU3). Specifically, Pritchard assumes a strong similarity between the classic Barn Façade Example and the Nero Example. I will argue that he is mistaken; the examples are in fact quite *different* in epistemically relevant ways, such that Nero's luck is compatible with his knowing. Consequently, proponents of lucky understanding saddle understanding with a kind of "non-epistemic" or "epistemically benign" luck.

To do this, I'll first cash out environmental luck's general structure with special attention to the Barn Façade Example. I'll then show that Pritchard's Nero Example exhibits a different structure, such that Nero's luck is compatible with his *knowing* that the house caught fire because of a faulty breaker box. Consequently, Pritchard's Nero Example is not a clear instance of understanding without explanatory knowledge.

To better clarify the structure of environmental luck, let's examine the modal make-up of the Barn Façade Example. Importantly, we can take Pritchard's Safety Principle (see Section 7.1) as our cue, for he explicitly argues that this principle explains why we should withhold knowledge in the Barn Façade Example. Importantly, this notion of luck does not discriminate between Gettier and environmental luck. To motivate that distinction, I offer the following:

Schema for Environmental Luck

An agent is environmentally lucky with respect to her belief that p if:
(1) In the actual world,
    a. The agent's belief that p is true, and
    b. The agent's belief that p is produced by one or more of her reliable cognitive abilities α and the fact that p; and
(2) In a nearby possible world, everything is the same as in the actual world except:
    a. The agent's belief that p is false, and
    b. The agent's belief that p is produced by α and the fact that p-*façade* (rather than the fact that p.)[6]

Here, cognitive abilities refer to those abilities that typically yield true beliefs – vision, inference, memory, and the like. For instance, in the Barn Façade Example, Bonnie's true belief is actually produced by her perceptual abilities and the barn, but a false belief could have easily been produced by those same abilities and a barn façade.

---

[6] Here, I profess agnosticism concerning any deep metaphysical issues about whether it is facts or something else (e.g., events) that produce beliefs, or whether production is reducible to a more rigorously analyzed concept (e.g., causation). The reader may fill in or substitute these variables as she sees fit. I also stipulate that α captures Pritchard's requirement that the belief be formed in the "same way" in both the actual and nearby possible world.

An essential feature of environmental luck is the notion of *p-façades*. While this is an admittedly loose concept, the crucial bit for my argument is that *p-façades* play the same role in producing the subject's belief that *p* in the nearby possible world as the truth-makers of *p* do in the actual world. For instance, the barn façade in the field plays the same role in producing Bonnie's belief in the nearby possible world as the barn in the field plays in producing that same belief in actuality. As I'll now argue, this is the crucial difference between the Barn Façade and Nero Examples for, if we were to follow the template for environmental luck presented earlier, the following would result:

Strict Nero Example
(1)   In the actual world,
      a.   Nero's belief that *the house's faulty breaker box explains why it caught fire* is true; and
      b.   Nero's belief that *the house's faulty breaker box explains why it caught fire* is produced by Nero's reliable ability to identify experts and the fact that the house's faulty breaker box explains why it caught fire.
(2)   In a nearby possible world, everything is the same as the actual world, except:
      a.   Nero's belief that *the house's faulty breaker box explains why it caught fire* is false; and
      b.   Nero's belief that *the house's faulty breaker box explains why it caught fire* is produced by Nero's reliable ability to identify experts and the fact that *something like* a faulty breaker box explains why the house caught fire.[7]

By contrast, Pritchard's Nero Example is naturally interpreted as follows:

Pritchard's Nero Example
(1)   In the actual world,
      a.   Nero's belief that *the house's faulty breaker box explains why it caught fire* is true; and
      b.   Nero's belief that *the house's faulty breaker box explains why it caught fire* is produced by Nero's reliable ability to identify experts and the fact that **he consulted the real firefighter**.

---

[7] Throughout this chapter, my talk reflects an "ontic" conception of explanation; i.e., that explanations are facts in the world. One could replace this with the other dominant view in which explanations are sets of true propositions, with the only loss being in economy of prose. For more on these distinctions, see Salmon (1989).

Table 7.1: *Differences Between the Barn Façade, Strict Nero, and Pritchard's Nero Examples*

	Belief	Fact	Relevant Contrast
Barn Façade	A barn is in the field.	A barn is in the field.	A barn (*rather than a barn façade*) is in the field.
Strict Nero	A faulty breaker box explains why the house caught fire.	A faulty breaker box explains why the house caught fire.	A faulty breaker box (*rather than a faulty-breaker-box-façade*) explains why the house caught fire.
Pritchard's Nero	A faulty breaker box explains why the house caught fire.	Not specified, though presumably the same as the Strict Nero Example	A real firefighter (*rather than a partygoer*) provides the testimony.

(2)   In a nearby possible world, everything is the same as the actual world except:
   a.   Nero's belief that (e.g.) **arson** *explains why the house caught fire* is false; and
   b.   Nero's belief that **arson** *explains why the house caught fire* is produced by Nero's reliable ability to identify experts and the fact that **he consulted a partygoer.**

Here, the boldface terms indicate the major points of difference between the Strict Nero Example and Pritchard's Nero Example: namely their *p-façades*. Table 7.1 summarizes the relevant similarities and differences.

Clearly, the Strict Nero Example is more faithful to the structure of the Barn Façade Example than Pritchard's Nero Example. Even if we grant that the truth-maker in the Strict Nero Example and Pritchard's Nero Example (i.e., the second column) are the same, the latter still trades in very different counterfactual claims, as is evidenced by the contrast it invokes (third column): Nero could have very easily consulted a different person, and that person would have offered false testimony. By contrast, fidelity to the structure of the Barn Façade Example suggests that Nero would have been indifferent to subtle explanatory differences (e.g., he might have confused a faulty breaker box with another explanation of the fire).

Disanalogies per se are harmless. However, in this case, these differences undercut the claim that the Barn Façade and Nero Examples are similar in *epistemically relevant* ways, which is what ELU3 states, and precisely what Pritchard needs for his argument to fly. Specifically, Pritchard's Nero Example seems to be flirting with an epistemically innocuous form of luck that arises when one could have easily received different evidence. For instance, consider the following:

Jesse James Example

> Jesse James robs a bank and is just about to make his grand escape when his mask accidentally slips off. A bystander, Byron, gets a clear view of James's face at this moment, and. owing to the latter's infamy, Byron immediately comes to believe that Jesse James robbed the bank. However, had Byron been standing at a slightly different location, he would not have formed a belief about the robber's identity.[8]

Now, despite the fact that Byron could have easily been standing at an angle where he did not see Jesse James's face, Byron knows who robbed the bank. Here is a more mundane example:

> Byron has an important meeting this morning, but is pressed for time and has misplaced his wallet. He has no clue where his wallet could be, and does not have time to scour his house. So, arbitrarily, the first place he checks is one of several seat cushions in his living room, where he finds his wallet. Had he checked any other location in the house, he would have rushed to work without his wallet.

Now, despite the fact that Byron could have easily checked somewhere else and not found his wallet, Byron knows that his wallet was underneath the seat cushion. Pritchard (2005, 137) treats these as examples of "benign" epistemic luck. More generally, he takes a whole class of examples of this sort to be compatible with knowledge:

Evidential Luck

> It is lucky that the agent acquires the evidence that she has in favor of her belief. (Pritchard 2005, 136)

Returning to Pritchard's Nero Example, a person's testimonial evidence is nothing more than the propositions she takes others to be asserting. Hence, in Pritchard's Nero Example, the real firefighter offers different testimonial evidence than the partygoer that Nero could have easily asked. Thus, Pritchard's Nero Example doesn't involve environmental luck; only

---

[8] Paraphrased from Nozick (1981).

evidential luck. Since evidential luck is compatible with knowledge, Nero may well *know* the explanation of the house fire. Thus, ELU3 is highly suspect.

Indeed, this speaks to a broader methodological problem in debates about the Luck Questions. In my travails with luck compatibilists, I've been struck by how few of their thought experiments fit the schema for environmental luck. Perhaps some kind of epistemic luck is not captured by the models I've set up, and epistemologists can intuit when this luck obtains. Let me profess that I (and perhaps the other incompatibilists mentioned earlier) lack these intuitions. To that end, I will make the possible-worlds models informing my thinking explicit and precise when discussing lucky understanding hereafter. Should these models clash with your intuitions, ask yourself the following: how would you set up your possible worlds differently from mine? Would your set-up require that the protagonist of your thought experiment could have easily deployed different evidence, cognitive abilities, or belief-forming processes? If you answer the latter question affirmatively, there's a good chance that you're trading in epistemically benign luck. Alternatively, you may be invoking a different anti-luck condition. However, note that at least as far as *Pritchard* is concerned, he has laid his cards on the table – it's safety or bust! – so this is not an option.

7.3.   Nouveau Incompatibilism

Let's take stock. Classic luck compatibilism fails on two fronts. First, it does not account for degrees of understanding. In other words, *if* updated luck incompatibilism is true, then the luck-driven objections to the received view dissolve. This is just to say that we should update our Luck Question. Second, classic luck compatibilism rests on examples that trade in epistemically innocuous luck. Hence, going forward, we'll want to make sure that we tighten up the modal structure of the examples so that we're actually testing for environmental luck. Specifically, we'll want to make sure that such examples are isomorphic to the Strict Nero Example described in Section 7.2.3.

I'll now argue that once we take these two points on board, we see that lucky understanding is best construed along precisely the lines suggested at the end of Section 7.2.2: a mere way station to better, safer understanding. However, it's going to take me some time to get there – you'll have to wait until Section 7.5.

Furthermore, I think that luck compatibilists have gotten something right: *merely* increasing safety does not (or, at the very least, need not) improve one's understanding. Where they've gone wrong is taking this to be a deep point about understanding's relationship to luck. After all, this is only a problem for those who think that understanding-why is *categorical* and equivalent to *mere* knowing-why. By contrast, those of us who hold that understanding *improves* as it more closely approximates *scientific* knowledge-why have ample resources for accommodating this insight. Specifically, scientific practice and, by implication, the EKS Model suggest that scientists go to great lengths to make sure that their understanding is safe and sound. This indicates that understanding improves to the extent that one's belief is not lucky because it emulates scientific practice.

I will develop this idea by motivating and defending the following:

Scientific Safety Argument

SS1.   If, according to scientific practice, x improves understanding of why p, then, if S_1 satisfies x and S_2 does not, then S_1 understands why p better than S_2.

SS2.   According to scientific practice, true beliefs about explanations that are safer because of better scientific explanatory evaluation improve understanding of why p.[9]

SS3.   ∴ *Scientific Luck Incompatibilism: Ceteris paribus*, if two people both have a true belief in the same explanation of p, then the person with the safer belief because of better scientific explanatory evaluation has a better understanding of why p. (SS1, SS2)

Note that the conclusion of this argument, scientific luck incompatibilism, is not the same as the updated luck incompatibilism presented in Section 7.1 for it requires that the luck be conquered through scientific explanatory evaluation (SEEing). I compare these two doctrines in Section 7.5.

For now, let me briefly discuss SS1. The guiding idea here is that we appeal to science when we look to exemplary instances of understanding. For the EKS Model, this idea is enshrined in the Scientific Knowledge Principle. Because I am restricting my analysis to explanatory understanding of empirical phenomena, note that putative counterexamples from non-empirical domains (e.g., mathematics) are not germane.

However, one might wonder if there's something like "everyday understanding" of empirical phenomena that isn't beholden to norms of

---

[9] There are implicit *ceteris paribus* clauses in the two premises. I leave them implicit for ease of presentation.

scientific appraisal. We should resist this idea for two reasons. First, for the sake of parsimony, it's preferable to have a single account of understanding that spans both scientific and everyday contexts. Second, even in "everyday" contexts, things like considering plausible alternative explanations, comparing them on the basis of good evidence, and forming beliefs accordingly are marks of more sophisticated understanding. Since these are the core features of scientific understanding, it's hard to see why scientific and everyday understanding should differ in kind – rather than in degree only. Hence, I will take SS1 as given hereafter.

I devote the lion's share of my efforts to clarifying and defending SS2. Specifically, while the next section grounds this claim in scientific practice, this section provides "armchair-epistemological" reasons for this premise. It will be instructive to begin with a simple example that we can use as a touchstone in discussing the various relationships among safety, scientific explanatory evaluation, and understanding:

Fiona and Carmen Example

Fiona is a firefighter who does an exemplary job in collecting all of the evidence that could be extracted from the embers of Nero's house. On the basis of this evidence, she comes to believe that a faulty breaker box is the root cause of the fire, and this is in fact true. However, at the moment of the fire, the breaker box malfunctioned at the same time that the grounding wire shorted. The two events were causally independent, and the shorted grounding wire did not actually cause the fire. But, had the shorted grounding wire (rather than the faulty breaker box) caused the fire, Fiona would have discovered exactly the same evidence and still believed that the breaker box's malfunctioning explains why Nero's house caught fire. By contrast, one of Fiona's colleagues, Carmen, considers both the breaker box and the grounding wire explanation. Moreover, she has all of Fiona's evidence, plus she takes some additional voltmeter readings at different points in the house, and these readings allow her to eliminate the grounding wire explanation.

Both Fiona and Carmen believe the same, true explanation: the house burned down because of the faulty breaker box. However, Carmen's belief is safer than Fiona's because she has engaged in better scientific explanatory evaluation. Because of this, she also understands why the house burned down better than Fiona.

While all of this bodes well for SS2, it is steeped in intuitions that others may not share. I'll now show that we can deliver the same result through more rigorous means. First, unlike Pritchard's Nero Example, Fiona's situation is very tightly patterned after the Barn Façade Example.

Consequently, we avoid the risk of conflating environmental and evidential luck. Note that instances of the Strict Nero Example (see Section 7.2.3)
give us clear examples of unsafe understanding of this sort. Specifically, in
Fiona's situation, the shorted grounding wire is playing the same role as a
barn façade in Goldman's example. Provisionally, let's call it an "explanation façade." Thus, Fiona is environmentally lucky with respect to her
belief that *the faulty breaker box explains the house's burning down* because:

Fı. In the actual world,

    a. Fiona's belief that *the faulty breaker box explains the house's burning
       down* is true, and

    b. Fiona's belief that *the faulty breaker box explains the house's burning
       down* is produced by her explanatory evaluation and the fact that *the
       faulty breaker box explains the house's burning down*; and

F2. In a nearby world, everything is the same, except:

    a. Fiona's belief that *the faulty breaker box explains the house's burning* is
       false, and

    b. Fiona's belief that *the faulty breaker box explains the house's burning
       down* is produced by her explanatory evaluation and the fact that
       the shorted grounding wire explains the house's burning down (rather
       than the fact that *the faulty breaker box explains the house's burning
       down.*)

Thus, we now have a clear instance of an unsafe belief in an explanation;
i.e., an instance of environmentally lucky understanding that fits the
aforementioned schema.

In this example, I've treated Fiona's explanatory evaluation as the
relevant ability. Here, I'm giving a nod to scientific explanatory evaluation (SEEing), which we've discussed in earlier chapters. Recall that
SEEing involves three core features: considering plausible potential
explanations of a given phenomenon, comparing those potential explanations by way of evidence and other explanatory considerations, and
forming beliefs on the basis of those comparisons. Fiona has considered
the breaker box explanation, found evidence for it, and formed her belief
accordingly.

Next, we need to show that Carmen's belief is safer than Fiona's. The
easiest way to see this is by comparing the modal structure of Fiona's belief
(Fı and F2) to Carmen's (differences in bold):

Cı. In the actual world,

    a. **Carmen's** belief that *the faulty breaker box explains the house's burning
       down* is true, and

    b. **Carmen's** belief that *the faulty breaker box explains the house's burning down* is produced by her use of the same explanatory evaluations that Fiona used, **plus the consideration of the grounding wire hypothesis and its elimination by the voltmeter readings**, and the fact that *the faulty breaker box explains the house's burning down*; and

   C2. In a nearby world, everything is the same except:

    a. **Carmen's** belief that *the faulty breaker box explains the house's burning* is false, and

    b. **Carmen's** belief that *the faulty breaker box explains the house's burning down* is produced by her use of the same explanatory evaluations that Fiona used – **but not the consideration of the grounding wire hypothesis and its elimination by the voltmeter readings!** – and the fact that *the shorted grounding wire explains the house's burning down* (rather than the fact that *the faulty breaker box explains the house's burning down.*)

Here, I'm assuming that the actual world and nearby possible world are the same for both Fiona and Carmen. Carmen's belief is then safer than Fiona's in two ways, both of which concern the differences between Fiona and Carmen's fates in the nearby possible world. First, although the possible world is nearby, it's not a world in which Carmen forms her belief in the *same* way as she does in the actual world. Thus, per Pritchard's Safety Principle, this possible world cannot render Carmen's belief unsafe, though this world does make Fiona's belief unsafe.[10] Second, because Carmen's belief-forming process is a more demanding belief-forming process than Fiona's, the closest possible world that begets an unsafe belief for Carmen will be further away from the actual world than the nearest possible world that renders Fiona's belief unsafe.

More generally, it suffices for my purposes that if two agents, S_1 and S_2, have beliefs and methods of explanatory evaluation that are isomorphic to Carmen and Fiona's, respectively, then S_1's belief is *safer* than S_2's. Indeed, we can see that small tweaks to Pritchard's Safety Principle yield a general account of comparative safety that is consistent with the preceding discussion:

    S_1's belief that *q explains why p* is safer than S_2's *iff*

   1. *Ceteris paribus*, S_1's belief that *q explains why p* continues to be true in more nearby possible worlds in which S_1 continues to form her belief

---

[10] Alternatively, one might think that, in this nearby possible world, Carmen has a *true* belief that the grounding wire caused the fire and that she formed her belief using her superior SEEing. The same general point would apply: this possible world would still not render Carmen's belief unsafe.

about *q explains p* in the same way as she did in the actual world, in comparison to corresponding worlds for S_2; or

2. *Ceteris paribus*, the closest possible world in which S_1's belief that *q explains why p* no longer continues to be true, but in which S_1 also continues to form her belief about the target proposition in the same way as the actual world, is farther away from the actual world than the corresponding world for S_2.

The *ceteris paribus* clauses are intended to capture the idea that we hold one of these two disjuncts fixed when assessing the other.

While we've shown that Carmen's belief is safer than Fiona's, we've yet to show that it's *because* of Carmen's *better* SEEing. To SEE better (hereafter: to "outSEE") is to form one's belief in an explanation on the basis of two different improvements. First, one may have considered and compared more potential explanations. To keep things simple, I will assume that if one person outSEEs another in this way, the latter's considerations and comparisons are a proper subset of the former's. For instance, whereas Fiona considers only the breaker box explanation and gathers evidence that supports this hypothesis, Carmen considers this explanation *and* the grounding wire explanation, and uses not only the same evidence that Fiona gathered in support of the former, but gathers further voltmeter readings to rule out the latter.

Second, one may have used more decisive evidence in one's comparisons (i.e., evidence that more strongly favors certain explanatory commitments). Of course, the voltmeter readings do just this. Here and throughout, I assume that SEEing entails that all such considerations and comparisons are in accordance with the relevant methodological norms of the domain. Hence, outSEEing can't be achieved through merely capricious considerations or comparisons.

Moreover, since the only difference between Carmen and Fiona is that the former outSEEs the latter, this explains the greater safety of the former's belief. It's in this sense that Carmen's safer belief is *because* of her better scientific explanatory evaluation. However, there is a deeper conceptual link between outSEEing and safer belief. A potential explanation is an explanation that is true in some possible world. In scientific contexts, a *plausible* potential explanation would thus appear to be an explanation that is true in some *nearby* possible world. Hence, when more plausible potential explanations are considered, more alternatives that could have easily explained the phenomenon of interest are considered. To compare explanations of this sort – at least in scientifically respectable ways – is to have good evidence as to how these explanations relate to the actual explanation. (In some cases, this relationship is one of identity; in other cases, preclusion; in still other cases, complementarity; etc.) Hence, when a person considers and compares more explanations – when she outSEEs another person or outSEEs her past self – her belief-forming process will be incompatible with more explanation façades. Hence, outSEEing goes hand in hand with safer belief.

Thus far, I've argued that it's possible to make one's explanatory commitments safer through better scientific explanatory evaluation. But why should we think that this improves one's understanding? Here, I'll lean on the dialectical space in contemporary discussions about understanding. Take your favorite understanding-boosting feature; by virtue of her better scientific explanatory evaluation, Carmen has more of it than Fiona. For instance,

- Carmen grasps more "coherence-making" relationships than Fiona: she grasps how the breaker box and grounding wire explanations relate to each other and to the voltmeter readings.
- Carmen can answer more what-if-things-had-been-different questions than Fiona: she can correctly predict how the voltmeter readings would have been different had the grounding wire (rather than the breaker box) caused the fire.
- Carmen has more true beliefs about the fire; e.g., that *even if the breaker box hadn't malfunctioned, the house would not have caught fire because of the shorted grounding wire*.
- Carmen's belief is a greater cognitive achievement than Fiona's; she has ruled out an especially thorny alternative explanation of the fire that Fiona has not.[11]

To repeat, the only difference between Carmen and Fiona is the former's outSEEing the latter via the consideration of the grounding wire explanation and its elimination on the basis of the voltmeter readings. Hence, Carmen's outSEEing explains her superior understanding.

To summarize, I've shown that better understanding results from safety earned through better scientific explanatory evaluation. The latter amounts to considering more plausible alternatives to the actual explanation – "explanation façades" – and having the ability to rule them out on the basis of sound evidence and scientific methods. This is the heart and soul of the Scientific Safety Argument, SS2. Scientific luck incompatibilism

---

[11] These bullet-points have been elaborated in Chapters 3 and 4.

follows: as we evaluate explanations more scientifically, our beliefs become safer, and our understanding improves.

7.4. Argumentum ex Scientia

However, one may worry that this is a hyper-idealized principle that only works with contrived thought experiments. To that end, I want to make the general conceptual link between these ideas and scientific practice more explicit (Section 7.4.1) and also illustrate how they work in an episode of science where it's quite clear that our understanding improved (Section 7.4.2).

7.4.1. *Understanding and Experimental Design*

Shorn of possible-worlds jargon, the underlying feature of lucky understanding is this: something other than the accepted hypothesis could easily explain the phenomenon of interest. What would happen if scientists discovered that they were in this situation? In slogan form: they would run an experiment and *control* for this explanation façade. However, they would call these façades "confounds." I will now argue that the practice of SEEing one's way to safer beliefs is deeply interwoven with the methodology of controlled experiments. Since such experiments are the standard-bearers for how scientists come to believe explanatory hypotheses, we have good reason to think that the ideas developed in Section 7.3 are not the stuff of make-believe, but track very closely with the norms implicit in scientific practice.

To begin, let's clarify what's meant by "controlling for a confound." Here, I'll avail myself to Woodward's account of an ideal intervention for useful details. Let I, X, and Y be variables. Then, according to Woodward, an ideal intervention I on X with respect to Y is a change in the value of X that changes Y, if at all, *only via* the change in X. More precisely:

(IV)

> I is an *intervention variable* on X with respect to Y if and only if I meets the following conditions:
> I1. I causes X.
> I2. I acts as a switch for all the other variables $[V]$ that cause X. That is, certain values of I are such that when I attains those values, X ceases to depend on the value of other variables that cause X and instead depends only on the value taken by I.

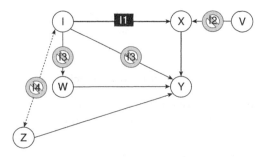

Figure 7.1: Conditions for Interventions.
The solid arrows indicate causal relations; the dotted line, a correlation; the black
box, the requirement I₁; and the gray, encircled strikethroughs, the restrictions I2–I4
discussed in the text.

I3. Any directed path from *I* to *Y* goes through *X*. That is, *I* does not directly
cause *Y* and is not a cause of any causes [*W*] of *Y* that are distinct from *X*
except, of course, for those causes of *Y*, if any, that are built into *I-X-Y*
connection itself: that is, except for (a) any causes of *Y* that are effects of
X (i.e., variables that are causally between *X* and *Y*) and (b) any causes of
Y that are between *I* and have no effect on *Y* independently of *X*.

I4. *I* is (statistically) independent of any variable *Z* that causes *Y* and that is
on a directed path that does not go through *X* (Woodward 2003, 98).

If *I* intervenes on *X* with respect to *Y*, then *I*, *V*, *W*, and *Z*, have been
controlled for (see Figure 7.1.)

Note that interventions of this sort are tightly connected to SEEing. It is
very difficult to design an experiment involving an ideal intervention that
controls for a given variable without having first *considered* that variable.
Moreover, an intervention is a way of *comparing* the effects of different
independent variables (*I*, *V*, *W*, *X*, *Z*) on the dependent variable (*Y*). When
one draws the right conclusion from such an experiment, one then *forms* a
belief on the basis of these prior considerations and comparisons. Hence, to
make an inference from an experiment is a kind of SEEing.

From this, we can also see why safety is an irresistible feature of
scientific practice. Once again, consider Fiona, whose belief in the
breaker box explanation is unsafe. For her, the crucial problem is
that she lacks the kind of evidence that would allow her to rule out
the possibility that the shorted grounding wire rather than the faulty
breaker box explains the house's burning down. Consequently, Fiona
fails to have evidence that I3 is satisfied: there is a causal path to the
house's burning down (*Y*) that can bypass the faulty breaker box (*X*) by

going through the shorted grounding wire (*W*).[12] By contrast, Carmen can rule out this possibility because of her voltmeter readings. Nor is this example idiosyncratic. It's in this sense that safety is a well-nigh inevitable feature of scientific explanatory evaluation: by controlling for more variables, scientists make their belief-forming processes (i.e., their explanatory evaluations) more demanding. This, in turn, makes their hypothesis space a safer place.

Nor is I3 the only way that this can be done. I1 guarantees that our independent variable *X* is explanatorily relevant to *Y*, and conditions I2–I4 make the possibility of *I, V, W*, or *Z* explaining *Y* remote. For instance, if I2 fails to obtain, then it is unclear whether a change in *Y* is explained by either a change in *X* or a change in *V*. Parallel points apply to I4 and *Z*. Thus, these conditions prohibit plausible rival explanations from easily being true. Hence, the notion of a controlled experiment – the paradigmatic scientific method for establishing explanatory hypotheses – has safety built into it.

For the remainder of this chapter, I assume that when scientists control for confounds, then those confounds are playing the role of *V, W*, or *Z* in the definition of ideal interventions presented earlier. Much of science – particularly the social sciences – falls well short of this ideal. Nevertheless, progress in these fields is often measured by how well they approximate these ideal interventions.[13] For instance, Carmen's voltmeter readings provide her with evidence that I3 has been satisfied, even if it's a stretch to claim that she "intervened" on the breaker box with the voltmeter. As Woodward (2003, 114) puts it, these conditions provide a "regulative ideal."

Importantly, controlling for more confounds is better than controlling for fewer (at least so long as the confounds are plausible alternatives). For instance, several authors have argued that explanations that hold under a wide range of conditions are preferable to those that do not. Depending on the author, this explanatory virtue is called scope (Lipton 2004), power (Lycan 2002), consilience (Thagard 1978), stability (Mitchell 1997, 2000), or resiliency (Skyrms 1980).[14] This also is a sign that ruling out alternatives is an important feature of our practices of explanatory evaluation. Essentially, more powerful explanations of this sort gain additional

---

[12] This, of course, is not an *experimental* intervention; more on this later.

[13] For a more thorough discussion of this, see Shadish, Cook, and Campbell (2001).

[14] This is not to say that these concepts are equivalent, only that they all seem to capture the idea that good explanatory generalizations hold under a wide variety of conditions.

justification by the evidence that they can explain and their competitors cannot.

Hence, if the practice of preferring these kinds of explanations is reliable, then safety via SEEing is important. Woodward's view on this issue is that explanations are frequently judged superior when they involve generalizations that are *invariant under more interventions.*[15] In other words, explanatory generalizations can be expressed as equations in which Y is a function of X, and these generalizations have greater explanatory power when they cannot easily be perturbed; i.e., they hold under a larger class of interventions. In this case, not only does each intervention provide the means of ruling out certain alternative explanations, but explanations that invoke generalizations that hold under a larger class of interventions are also deemed superior. Hence, the greater the invariance of an explanatory generalization, the greater resources it provides for comparing alternative explanations.[16] The key point here is that scientists tend to find these explanations to be especially valuable.

Thus, to summarize, experimental inference is the paradigmatic scientific method for forming beliefs about explanations. Experimental inference entails safety through SEEing. In particular, scientists consider confounds and control for them using interventions. Moreover, if an explanation remains invariant as more of these confounds are considered and controlled for, scientists' estimate of that explanation increases. Hence, it appears that explanations that are especially amenable to safety through SEEing are highly valued in scientific practice. What might scientists call this value that they attach to these explanations? "Understanding" seems an apt term.

7.4.2. Peptic Ulcers

In the preceding discussion, I offered fairly general reasons to think that SS2 is true of scientific practice. However, there's plenty of room between the rarefied air of methodological reflection and the trenches of actual science. To that end, I want to offer a more detailed account of how these

---

[15] My use of Woodward is only meant to vividly illustrate that excluding rival explanations is an important aspect of explanatory evaluation, but any epistemology of explanation says this much. Hence, I submit that careful readings of, e.g., Hempel (1966), Kitcher (1989), Lipton (2004), Popper (1963), and Thagard (1978) will furnish very similar support for SS2 of this argument. Indeed, insofar as explanatory evaluation is a species of hypothesis testing, virtually every theory of confirmation also requires that plausible alternatives be ruled out.

[16] For more on invariance, see Hitchcock and Woodward (2003), Woodward (2003, ch. 6), and Woodward and Hitchcock (2003).

considerations bear on an actual episode in the history of science: the early history of the bacterial theory of ulcers.

Peptic ulcers are sores that develop in the stomach (gastric ulcers) or in the duodenum (duodenal ulcers). Through the 1970s, biomedical scientists held that excess acidity in the stomach causes these ulcers. Starting in the 1970s, antacids were used as effective relief from peptic ulcers, although they did not cure ulcers. Furthermore, it was assumed that bacteria could not survive in the stomach's acidic environment. However, as first conjectured in 1983 by Australian physicians Barry Marshall and Robin Warren, biomedical scientists now hold that bacteria cause peptic ulcers. Consequently, antibiotics are now used to treat them, with much greater success.

This much is clear: discovering that bacteria cause ulcers improved our understanding of why some people suffer from peptic ulcers. However, I want to focus on a much narrower time-slice in this discovery: namely, the transition of the bacterial theory of ulcers from a suggestive hypothesis in 1983 to its earliest empirical successes in 1984. To that end, I'll focus on one of Marshall and Warren's (1984) earliest published articles in support of the bacterial theory.

In that study, they discovered the bacteria that would later be called *Helicobacter pylori* in the stomach biopsies of several people with gastritis and inferred that the bacteria explains the gastritis. Consonant with the first and second features of scientific explanatory evaluation, they used techniques and evidence designed to eliminate several alternative explanations of why the patients have gastritis or how the bacteria entered the patients' systems; e.g.,

> Where possible patients completed a clinical questionnaire designed to detect a source of infection or show any relationship with "known" causes of gastritis or *Campylobacter* infection, rather than give a detailed account of each patient's history. The emphasis was on animal contact, travel, diet, dental hygiene, and drugs, rather than symptoms. (1984, 1311)

Hence, here we can see them using the evidence from the questionnaire to control for animal contact, travel, and the like as potential confounds.

However, these were far from the only controls. Marshall and Warren required patients to fast at least four hours before the endoscopy, used certain stains (e.g., hematoxylin and eosin (H&E), Warthrin-Starry silver, Gram), cultured the samples, and had their results independently coded, all done to rule out certain confounds.

Moreover, our safety requirement on understanding provides a plausible *raison d'être* for why scientists undertake these measures: to make sure that their explanations could not easily have been false. In modalease, they are setting up an experimental situation such that, in all nearby possible worlds, the presence of bacteria explains why the patients have gastritis, just as our discussion of Woodward suggests.

For instance, Marshall and Warren found evidence that only the bacteria hypothesis explained: namely, various endoscopy, histopathology, and microbiology results. With the aforementioned questionnaires, they discovered the following:

> The only symptom which correlated with gastritis or bacteria was "burping" which was more common in patients with bacteria ($p = 0.03$) or gastritis ($p = 0.007$). This association remained when patients with peptic ulcer were excluded. None of the other questionnaire responses showed any relationship to the presence of gastric bacteria or gastritis.

In other contexts, the use of significance testing provides further evidence that safe explanatory evaluation figures prominently in the scientific practices that afford us understanding because a low p-value indicates that the correlation between explanans and explanandum could not easily have been a fluke (i.e., the null hypothesis is not true in any nearby possible worlds.) For instance, the endoscopy results indicated a very close correlation between ulcers and bacteria ($p = 0.0002$).

We see more evidence of SEEing guiding Marshall and Warren's (1984, 1312) study when we turn to the histopathological tests for their explanation:

> Gastritis could usually be graded with confidence at low magnification. There was some difficulty with about 25 cases where the changes were mild or the specimens were small, superficial, or distorted. To ensure that gradings were reliable, single H & E sections from the last 40 cases were examined "blind" by another pathologist who agreed with the presence or absence of gastritis in 36 cases (90%), and gave an identical grading in 32.

Thus, once again, the scientists achieved understanding only when they created a "safe space" for their explanations.

Regarding the third feature of SEEing, belief-formation, Marshall and Warren (1984, 1314) first discuss the failure of other explanations of gastritis and ulcers, and then assert the following:

> We know of no other disease state where, in the absence of complicating factors such as ulceration (table IV), bacteria and PMNs [polymorphonuclear leucocytes, a telltale sign of gastritis] are so intimately related without the bacteria being pathogenic.

This is not quite an assertion that the presence of bacteria explains gastritis, but that is consonant with my idea that a doxastic state should be based on the explanatory comparisons. Since Marshall and Warren were offering a brand new explanation of gastritis, they might reasonably have thought that more explanatory evaluation was in order. Hence, their measured claims about this explanation are in line with my account of explanatory evaluation.

Importantly, this was just the beginning of this explanation's career; several subsequent observations and experiments precipitated this advance in our understanding of peptic ulcers. Marshall and Warren discovered that antibiotics cure peptic ulcers. Later, the previous consensus that the stomach was too acidic to host bacteria was flatly refuted because *H. pylori* was microscopically observed and was grown in laboratory cultures (Marshall et al. 1990). Also, several studies indicated higher rates of ulcer healing and lower rates of recurrence among ulcer sufferers in whom *H. pylori* was eradicated (e.g., Marshall et al. 1988). Parallel points about safety and SEEing's three features apply to these studies.

Now, here's the big point: why would Marshall and Warren (as well as subsequent ulcer researchers) engage in all of these controls and statistical tests? After all, they had already conjectured that bacteria cause ulcers in 1983, before running any of these tests. It nevertheless seems eminently reasonable to deny that they understood why people had ulcers in 1983, and this denial is warranted precisely *because* they were merely speculating. Our account of SEEing helps to illuminate this judgment. Mere speculation, at best, requires only consideration of an explanation. By contrast, experiments and their statistical scaffolds earn their keep in the process of explanatory comparison. Additionally, we've seen that these various experimental controls and statistical tests are consonant with the idea that we achieve understanding by considering and comparing alternative explanations with an eye toward forming explanatory commitments that could not easily have been false. Hence, it would appear that safety through SEEing was a crucial ingredient in advancing our understanding of ulcers.

7.5.   Conclusion

Let's take stock. First, the basic feature of lucky understanding is a salient possibility of something other than one's accepted hypothesis explaining the phenomenon of interest. Second, scientists respond to these possibilities by considering them and then comparing them – paradigmatically in the context of controlled experiments. Third, we have every reason to

believe that when these experiments are well designed and properly interpreted, there is a gain in understanding. So, with this, all of the elements of
SS2 are in place:

> SS2.   According to scientific practice, true beliefs about explanations that
> are safer because of better scientific explanatory evaluation improve
> understanding of why p (*ceteris paribus*).

When coupled with the plausibility of SS1, we now have good reason to
accept our EKS-infused version of updated luck incompatibilism:

> *Scientific Luck Incompatibilism: Ceteris paribus*, if two people both have a
> true belief in the same explanation of p, then the person with the safer
> belief **because of better scientific explanatory evaluation** has a better
> understanding of why p.

The boldface phrase indicates the central difference with a similar position
we staked out in Section 7.1, viz.:

> *Updated Luck Incompatibilism: Ceteris paribus*, if two people both have a
> true belief in the same explanation of p, then the person with the safer
> belief has a better understanding of why p.

Consequently, *avant la lettre*, it would appear that the luck compatibilists
have gotten the better of this exchange. Indeed, the preceding suggests that
we can be much more precise about when increased safety does not
improve our understanding. In particular, *updated* luck *compatibilism*
and *scientific* luck *incompatibilism* will both be true if:

> There are cases in which two otherwise-identical people both have a true
> belief in the same explanation of p – **and scientifically evaluate explana-**
> **tions of p in exactly the same way** – but the person with the safer belief
> does not have a better understanding of why p.

However, I want to suggest that, if this is the only way in which updated
luck compatibilists can make their case, it's a pyrrhic victory. In particular,
this is really no advance over the arguments for *classic* luck compatibilism
and faces a similar liability. Recall how the classic cases were handled: the
existence of lucky understanders is perfectly consistent with updated luck
incompatibilism, since we could always grant that this is an instance of
outright understanding while holding fast to the idea that safer understanding is *better*. That point doesn't change simply because we add the
boldfaced riders about SEEing.

To see this, imagine a third firefighter, Tommy, whose epistemic situation is somewhere in between Fiona's and Carmen's. Specifically, Tommy

evaluates the evidence and explanations in exactly the same way as Fiona, but also asks Carmen for her opinion. Furthermore, Tommy takes Carmen's testimony to provide independent and additional corroboration for his belief that the faulty breaker box caused the fire, but he has no idea about the voltmeter readings that led Carmen to her assertion about the fire's causes. Hence, Tommy's understanding is safer than Fiona's but not because his SEEing is more circumspect. Let's grant that Tommy's understanding is no better than Fiona's. This clearly satisfies the strictures of updated luck compatibilism.

However, at this point, it's already clear why the game is up: *Carmen* has better understanding than *either Tommy or Fiona*. Furthermore, she does so in accordance with scientific luck incompatibilism. This shows that we can still answer the Updated Luck Question in precisely the way that the EKS Model suggests: understanding improves as it becomes less lucky, so long as that luck is overcome *scientifically*. It's in this sense that the victories for luck compatibilists are pyrrhic.

In conclusion, we have rehearsed the ins and outs of how understanding is alleged to either tolerate or resist luck. Arguments in favor of lucky understanding were found wanting, especially after we introduced the (now-mundane) idea that understanding admits of degrees. At the very least, these arguments are compatible with understanding improving as it becomes less lucky. Indeed, I have argued that such improvements paradigmatically involve explanations secured through scientific methods. In short, when our explanations are safer because we have vetted them scientifically, our understanding flourishes.

This also concludes a major arc of our dialectic. Whereas Chapters 4 and 5 argued that divorcing understanding from *explanation* provided insufficient motivation for fancy new theories of understanding, this chapter and its predecessor show that the defining features of *knowledge* are also woven into understanding's DNA. Degrees of understanding are measured by their proximity to scientific knowledge of an explanation. The epistemology of scientific explanation may not be the received view, but it is an epistemology of scientific explanation nonetheless.

7.6. Excursus: Gettier Luck

I have argued that the most developed account of lucky understanding, which focuses on environmental luck, does not effectively undermine the EKS Model. However, recall that in addition to environmental luck, *Gettier-style* luck also conflicts with knowledge. This was luck that intervened between a belief and its truth-maker. I'll briefly explain why the preceding arguments apply to Gettier-style luck.

While Pritchard rejects the idea that Gettier-lucky understanding is possible, others are willing to countenance it (Morris 2012; Rohwer 2014). Morris's (2012) arguments cast a very wide net and essentially entail that a true belief in an explanation suffices for understanding regardless of *any* kind of luck (Gettier, environmental, or otherwise.) As an example, he offers a version of the following:

Nellie Example

Nellie comes home to find her house in flames. When she asks a firefighter what caused the fire, he gives her the correct answer that it was a faulty breaker box. Unbeknownst to Nellie, the person she asked is one of several people dressed as firefighters en route to a costume party. Furthermore, the fake firefighter completely fabricated the story about the faulty wiring and could have very easily provided a false answer to Nellie.

While some take Nellie to lack understanding in this case, Morris (2012, 369) demurs. The argument looks quite similar to Pritchard's:

Gettier-Lucky Understanding Argument

GLU1. Nellie understands why the house burned down.
GLU2. If two beliefs are similar in all epistemically relevant respects, and if one of those beliefs is unsafe, then so is the other.
GLU3. Abel's belief in the Sheep Example and Nellie's belief in the Nellie Example are similar in all epistemically relevant respects.
GLU4. Abel's belief in the Sheep Example is Gettier-lucky.
GLU5. If a belief is Gettier-lucky, then it is unsafe.
GLU6. $\therefore$ *Classic Luck Compatibilism:* Understanding why p does not require a safe belief in any explanation of p. (GLU1–GLU5)

Now, as should be clear, Morris's first premise GLU1 and third premise GLU3 are subject to precisely the same objections Pritchard's ELU1 and ELU3 in Sections 7.2.2 and 7.2.3, respectively. Focusing only on the first, it's no surprise that outright understanding in a relatively low-standards context can be unsafe once we countenance the EKS Model. The main question is whether making it safer counts as an improvement. Hence, regardless of the luck involved, it's time to update our thoughts about luck and understanding.

Additionally, the account I gave in Sections 7.3 and 7.4 applies just as readily to Gettier-lucky understanding. To see this, let's look at the structure of Gettier-lucky understanding:

Schema for Gettier-Lucky Understanding

An agent is Gettier-lucky with respect to the belief that *q explains why p* if:

(1) In the actual world,
   a. the agent's belief that *q explains why p* is true, and
   b. the agent's belief that *q explains why p* is produced by the fact that *r explains why p* (rather than the fact that *q explains why p*.)
(2) In a nearby possible world, everything is the same as the actual world except:
   a. the agent's belief that *q explains why p* is false.

In Section 7.3, I noted that the underlying idea behind environmentally lucky understanding were "explanation façades," which are scenarios in which something other than one's accepted hypothesis could easily explain the phenomenon of interest. As should be clear, precisely the same problem looms here. Indeed, in Gettier-lucky understanding this possibility is even less remote than in environmentally lucky understanding for the confound or "explanation façade" obtains in the actual world. However, precisely because it's the same kind of problem, it admits of the same solution: outSEEing by way of ideal interventions. Hence, my discussion of environmentally lucky understanding readily applies to Gettier-lucky understanding.

CHAPTER 8

The Value of Understanding

"I wonder that knowledge should be preferred to right opinion – or why they should ever differ." Thus spoke Meno, the interlocutor in the eponymous Socratic dialogue. To use Socrates' example, why is knowledge about Larissa's whereabouts more valuable than a mere true belief to the same effect? After all, both will get you where you need to go. What further job is knowledge doing that isn't epistemic overkill?

Meno's question – whether knowledge is more valuable than true belief – is an interesting one. Indeed, since that fateful conversation in the agora, the increasingly byzantine analyses of anti-Gettier conditions make Meno's question even more pressing. What's so great about avoiding the epistemic mischief caused by barn façades, book-stealing doppelgangers, unusually positioned sheep, dishonest car-owning coworkers, the coins in an employee's pocket, and other curiosities that have kept epistemologists gainfully employed for the past fifty years? Why are true beliefs with *those* kinds of features worth having?

Despite its importance, the so-called Meno problem has only crept into the epistemological consciousness in the past decade or so. A prominent (though by no means unanimous) position has emerged: knowledge *isn't* valuable in the right kind of way, but understanding *is*. Indeed, the fountainhead of such value-driven epistemology – Kvanvig's (2003) *The Value of Knowledge and the Pursuit of Understanding* – adopts precisely this stance. Having said this, Kvanvig is by no means alone (Carter and Gordon 2014; Kvanvig 2003, 2009a, 2009b; Pritchard 2008, 2009a, 2010; Zagzebski 2001).

So, perhaps it is not the *nature* of understanding that makes it shiny and new. Perhaps, instead it is the *value* of understanding. Moreover, presumably the value of an epistemic status supervenes upon its nature. Thus, because I have argued that the differences between the nature of explanatory knowledge and understanding are modest, it is unclear how I can capture the unique value of understanding.

Of course, this assumes that understanding actually has a distinctive value. I will argue that it does not. This brings us to the last of the challenges to the received view mentioned in Chapter 1, the *General Value Question:*

• Do understanding and knowledge have the same kinds of epistemic value?

Contrary to Kvanvig and his sympathizers, I will argue that the General Value Question should be answered in the affirmative.

To defend this position, I will have to discuss two further questions concerning the value of understanding. The Explanation-Knowledge-Science (EKS) Model ably answers the first of these questions (Section 8.1). By contrast, we lack a good reason to ask – much less answer – the second question, which demands that we find differences between the *kinds* of value that attach to understanding and other, less demanding epistemic statuses (Section 8.2). Since knowledge fares no better, we should answer the General Value Question affirmatively: when compared to knowledge, there is nothing distinctive about understanding's value. I conclude both the chapter and the book with broader reflections about the nature and value of understanding (Section 8.3.)

8.1. The Basic Value Question

In order to answer the General Value Question, which seeks to ascertain any differences in the value of understanding and the value of knowledge, I must first answer a more elementary question:

• *Basic Value Question:* Why is an increase in understanding valuable?

Let's define the key terms here. An increase in understanding is simply when one has better understanding as prescribed by the EKS Model, i.e.:

(EKS1) S_1 understands why p better than S_2 if and only if:
   (A) *Ceteris paribus,* S_1 grasps p's explanatory nexus more completely than S_2; or
   (B) *Ceteris paribus,* S_1's grasp of p's explanatory nexus bears greater resemblance to scientific knowledge than S_2's.

Note that there are two kinds of evaluations going on with the Basic Value Question. Specifically, there are first-order, epistemological evaluations of agents' understanding (e.g., "Does one person understand why *p* *better* than another person? Does the agent have *ideal* understanding?") and second-order, axiological evaluations of agents' understanding ("Why is improving one's understanding in accordance with the EKS Model a *good* thing?"). The first-order evaluations are not at stake in any of the Value Questions; only the second-order evaluations are. Indeed, given the arguments in preceding chapters, the EKS Model is my final word on these first-order evaluations. I will flag second-order considerations by talking about *more valuable* understanding. By contrast, I will refer to first-order issues by talking about *better* understanding, *improvements* to our understanding, or *increases* in our understanding.

Additionally, I will be focusing on understanding's *epistemic* value. Undoubtedly, understanding has many practical benefits as well, but epistemologists have paid scant attention to these benefits. Consequently, subsequent references to "value" are elliptical for "epistemic value."

The EKS Model provides a relatively straightforward answer to the Basic Value Question (Section 8.1.1). Thus, understanding is valuable in *some* sense. In particular, we reap epistemic value as our understanding advances. It also furnishes an answer to several other important questions about understanding's value – including the General Value Question (Section 8.1.2).

8.1.1.   The Answer

Having clarified the Basic Value Question concerning understanding, let me follow a clear-cut strategy for answering it: we enumerate the different ways that understanding can increase or improve, and identify the epistemic value that accrues to each kind of improvement. We can tease these different kinds of emendations in our understanding from the EKS Model's two core principles. Four such improvements emerge.

First, recall the Nexus Principle:

> *Ceteris paribus,* S_1 understands why *p* better than S_2 if S_1 grasps *p*'s explanatory nexus more completely than S_2.

This suggests that one way of increasing understanding is to grasp more explanatory information. Taken neat, this grasping will be something in the neighborhood of the following:

- *Explanatory improvements:* true beliefs about a wider variety of correct explanations of *p*.

True belief enjoys a default status as a fundamental epistemic good (e.g., David 2001; Grimm 2009a; Lynch 2004, 2009). In what follows, I will assume that true belief is *a* fundamental epistemic good, but I make no further assumption that it is the *only* good of this kind.

Given these caveats, any explanatory improvement is more valuable in degree than that which falls short of it – even less demanding instances of understanding. For instance, suppose that you know that a car accident occurred both because of slippery roads and because of excessive speed, while I only know that it occurred because of excessive speed. Your understanding is better than mine. Since you are in possession of more (explanatorily relevant) truths than I am, your understanding is also more valuable. Hence, explanatory improvements are of fundamental epistemic value.

Indeed, this also provides a springboard for explaining why minimal understanding is more valuable than that which falls short of it. Consider a true belief *that p*. To minimally understand why *p* involves a true belief that *p*, plus a true belief about that which *explains p*.[1] Consequently, given the assumption that true beliefs are of (fundamental) epistemic value, there is a difference in degree here: one has (at least) two true beliefs when one understands why *p*, which needn't be the case when one merely has the true belief that *p*.

The preceding works fine given the contrast. But one might wonder why it's more valuable to have minimal understanding – a true belief that *q explains why p* – rather than, e.g., a true belief in the conjunction, *p and q*. Presumably, tackling this problem requires an account of the value of an explanation. While I think this is a perfectly good topic of philosophical discussion, here are two reasons that I won't be the person to explore it (at least not here).

First, the assumption that explanations are valuable seems widely shared among those working on the value of understanding-why. Hence, it's dialectical fair game. Second, a story about explanatory value should lean on some invariant feature or role that all explanations share. However, because I am an explanatory pluralist, no simple version of this story is forthcoming. Additionally, my agnosticism about explanatory realism further complicates that story. For instance, van Fraassen, our arch-anti-realist, construes explanatory value as entirely instrumental; realists are likely to view it as having some final value. So, to lean on the value of

---

[1] Here and throughout, I bracket the axiological implications about EKS2's reference to approximation.

explanation, these details will have to be worked out in a way that I can't possibly develop here.

Moving back to the big picture, we have only canvassed one of the four kinds of improvements in understanding. We get the other three from the Scientific Knowledge Principle:

> *Ceteris paribus*, S_1 understands why p better than S_2 if S_1's grasp of p's explanatory nexus bears greater resemblance to scientific knowledge than S_2's.

Given our account of scientific knowledge, such improvements should be characteristic of scientific explanatory evaluation (SEEing). This suggests our three remaining improvements:

- *Improved consideration:* more plausible alternative explanations are considered;
- *Improved comparison:* these alternative explanations are compared using better scientific evidence and methods; or
- *Scientific improvement:* beliefs about explanations are made safer because of these considerations and comparisons.

As should be clear, improved consideration and improved comparisons are largely a means to scientific improvements. Hence, when understanding witnesses improved consideration or improved comparison, its instrumental value increases in degree.

But if improved consideration and comparison are a means to scientific improvement, then it's incumbent upon us to identify the latter's value. Here we can lean on earlier work. The leading (if not the only) reason to think that *knowledge* is distinctively valuable comes from what is sometimes called "robust virtue epistemology" (Greco 2009; Sosa 1991, 2007; Zagzebski 1996). According to this view, knowledge is a true belief that is primarily creditable to an agent's cognitive abilities. For instance, perceptual knowledge is a true belief primarily creditable to one's perceptual capacities; inferential knowledge to one's inferential capacities; and so forth. So, robust virtue epistemologists hold that knowledge is distinctively valuable because it is a kind of achievement (i.e., a success that is primarily because of a person's ability).

Specifically, these virtue epistemologists hold that a knower has a kind of cognitive success – a true belief – primarily because of her cognitive abilities. Thus, for robust virtue epistemologists, knowledge is a cognitive achievement. Intuitively, achievements of all sorts are *finally* valuable in the sense that their value is not exhausted by their conduciveness to other good things. Thus, in broad outline, robust virtue epistemologists hold that knowledge is finally valuable because it is a kind of achievement.[2] Can this idea be repurposed for scientific improvements? Here, I will take some small liberties with an argument from Pritchard (2010, 80–84):

The Scientific Achievement Argument

SA1. Achievements are successes that are primarily because of ability, where the success in question either involves the overcoming of a significant obstacle or the exercise of a significant level of ability.

SA2. A scientific improvement is a cognitive success that is primarily because of cognitive ability, where the success in question either involves the overcoming of a significant obstacle or the exercise of a significant level of ability.

SA3. Achievements have final value.

SA4. ∴ Scientific improvements have final value (SA1–SA3).[3]

Let's discuss each premise in turn. Some successes because of ability are not achievements – or at least not very significant ones. For instance, intentionally raising one's arm is a success because of ability, but it would be hyperbolic to call it an achievement in most circumstances. SA1 addresses this worry by adding the further rider that the success must involve overcoming a significant obstacle or exercising a significant level of ability. For instance, raising one's arm may be an achievement if one is overcoming a serious obstacle (e.g., an injury) or exercising a significant skill (e.g., weight-lifting or contortionism) but is otherwise not an achievement. Furthermore, only these significant or "strong" achievements are liable to be finally valuable. Thus, SA1 lends credence to SA3.

What of SA2? As we saw in the previous chapter, scientific understanding frequently advances in accordance with the following principle:

Scientific Luck Incompatibilism: Ceteris paribus, if two people both have a true belief in the same explanation of *p*, then the person

---

[2] See Pritchard (2010) for criticisms of this proposal with respect to knowledge. This won't be our concern here.

[3] The aforementioned virtue epistemologists tend to have a broader notion of a cognitive achievement than the kind referred to in SA2. Consequently, they should also accept this argument.

with the safer belief because of better scientific explanatory evaluation has a better understanding of why p.[4]

This already furnishes a true belief that is creditable to an agent's ability to
engage in scientific explanatory evaluation. Will this always involve exercising a significant ability or overcoming a significant obstacle? Certainly
this will often be the case, since scientific research typically involves the
exercise of significant cognitive abilities, such as imagination in conceiving
of alternative explanations of a phenomenon, mastery of challenging background literature in considering other alternatives, learning and applying
methods by which to compare those alternatives, and keen inferential
acumen when forming beliefs about those explanations (see Section 3.2.3
for further details).

Finally, SA3 states that achievements are finally valuable. A state of
affairs is merely instrumentally valuable if and only if its value is exhausted
by its being a means to something else that is of value; otherwise, it has
some final value. Thus, cognitive achievements are finally valuable in the
sense that their value is not exhausted by their conduciveness to other good
things. Here, I can do no better than to quote Pritchard (2010, 30):

> Imagine, for example, that you are about to undertake a course of action
> designed to attain a certain outcome and that you are given the choice
> between merely being successful in what you set out to do, and being
> successful in such a way that you exhibit an achievement. Suppose further
> that it is stipulated in advance that there are no practical costs or benefits to
> choosing either way. Even so, wouldn't you prefer to exhibit an achieve-
> ment? And wouldn't you be right to do so? If that's correct, then this is
> strong evidence for the final value of achievements.

In other words, achievements are finally valuable because an outcome
exhibiting an achievement is preferable to an otherwise identical outcome
lacking in achievement.

At this point, I take the Scientific Achievement Argument to be on firm
ground. More importantly, I have discussed the four ways in which the
EKS Model entails that understanding can increase and have shown that
each provides an answer to the Basic Value Question. Understanding can
improve *explanatorily* when one has more true beliefs about correct explanations and true beliefs are of fundamental epistemic value. Understanding
can improve *scientifically* when one has safer beliefs as a result of engaging

---

[4] Interestingly, improvements in understanding that have this structure will be a species of what Turri
(2016) dubs "ample" or "super-safe" cognitive achievements. Turri offers his own reasons for
thinking that these ample achievements accrue special value. I won't pursue that here.

in better scientific explanatory evaluation. This improvement is finally valuable because it is an achievement. Additionally, we can sometimes enjoy boosts in the instrumental value in our understanding when our capacity to *consider* or *compare* explanations improves, even if we do not manage to form safe beliefs as a result. Hence, for any increase in understanding, we can show that it is more valuable than anything that falls short of it. The Basic Value Question is answered.

8.1.2. The Good News . . .

The Basic Value Question asks why increases in understanding are valuable. In other words, why is it a good thing when our understanding advances? We can answer this: because such advances involve amassing more true beliefs, accruing greater instrumental epistemic value, and exercising one's cognitive abilities so as to achieve some epistemically praiseworthy end. This has two important consequences concerning the value of understanding.

First, and most importantly, by answering the Basic Value Question, we can also answer the General Value Question. Recall that the latter asks if understanding and knowledge have the same kinds of epistemic value. As we've seen, understanding's epistemic value consists of the fundamental epistemic value of true belief, instrumental epistemic value, and the "achievement-value" championed by virtue epistemologists. Knowledge can possess all of these values, too. Hence, understanding and knowledge have the same kinds of value. So, the General Value Question admits of a positive answer.

To be sure, knowledge isn't always an achievement (Lackey 2007; Pritchard 2010). But, then again, neither is understanding, as I have argued in Chapter 3. In particular, minimal understanding need not be a cognitive achievement. Parallel points apply to instances of outright understanding that are only marginally better than minimal understanding.

One might also complain that we have assumed that explanation is valuable, and, while understanding-why necessarily enjoys this value, knowledge does not. The merits of the objection are hard to assess without a substantive account of explanation's epistemic value. Nevertheless, let me grant that only epistemic statuses that require explanation can lay rightful claim to this value. Since knowledge-why requires explanation, then something quite close to our answer lies in the wings: knowledge-why and understanding-why accrue the same values. Feel free to affix this caveat where relevant.

Indeed, I think this gentle tweak to my position rests on a very general methodological assumption needed to compare the value of different cognitive statuses. To see the importance of this assumption, suppose that we are trying to answer the General Value Question on the basis of the following comparison:

- Ideal understanding of something frivolous – say, why some tabloid celebrity is wearing a designer jumpsuit.
- Knowledge of something important – say, the exact measures needed to achieve world peace.

This contrast is too "noisy" to give us the insights we need to determine whether knowledge and understanding have the same kinds of value. The value of what is understood is so miniscule compared to that of what is known that it doesn't allow us to pinpoint anything about understanding or knowledge per se. The right kind of contrast would involve understanding and knowing the same thing. For instance, either we should compare understanding and knowledge of celebrity jumpsuits, or we should compare understanding and knowledge of world peace. What we shouldn't do is mix and match celebrity jumpsuits and world peace, as the original example did. So, by implication, we should be comparing understanding of why p to knowledge of why p. Since both require explanations, the objection is blunted.

Thus far, I've shown that answering the Basic Value Question suffices to answer the General Value Question. Recall, however, that a second point about epistemic value is in the offing. Specifically, my answer to the Basic Value Question allows me to account for the value of ideal and outright understanding. For instance, we can explain why ideal understanding is more valuable than anything that falls short of it. To see this, recall our definition of ideal understanding:

> S ideally understands why p if and only if it is impossible for anyone to understand why p better than S.

Ideal understanding will simply have more of the three values characteristic of increases in understanding. Similarly, recall our definition of outright understanding:

> "S understands why p" is true in context C if and only if S has minimal understanding and S approximates ideal understanding of why p closely enough in C.

For any instance of outright understanding, we will be able to account for why it is more valuable than that which falls short of it, as the former will furnish better understanding than the latter. As a result, my answer to the Basic Question accounts for why understanding garnered in more demanding contexts is of greater epistemic value than its counterpart in less demanding contexts. For example, suppose that both Kate and Hank understand why the sky is blue, but Kate understands better than Hank. Then, an answer to the Basic Question will tell us why Kate has something of greater epistemic value than Hank.

Finally, recall our concept of minimal understanding:

(EKS2) *S* has minimal understanding of why *p* if and only if, for some *q*, *S* believes that *q explains why p*, and *q explains why p* is approximately true.

For the reasons just canvassed, an answer to the Basic Question will explain why non-minimal outright understanding is more epistemically valuable than minimal understanding. (To see this, simply imagine that Hank has minimal understanding in the preceding example.)

Earlier, I also explained why minimal understanding is better than that which falls short of minimal understanding. It requires more true beliefs than a true belief that *p*, and it also partakes of explanatory value. Hence, we have a complete story about explaining why the value of any degree of understanding is more valuable than that which falls short of it.

A quick synopsis of the good news is in order: I have explained why improving our understanding is epistemically valuable and also why minimal understanding is better than true beliefs that fall short of it. This explanation also tells us why knowledge and understanding have the same kinds of epistemic value, and why ideal and outright understanding are more valuable than understanding that falls short of them.

8.2. The Distinctive Value Question

My answer to the General Value Question is affirmative: knowledge and understanding have the same kinds of epistemic value. This stands opposed to the epistemologists mentioned in the introduction, who held that understanding is valuable in a way that knowledge is not.

To appreciate the motivations of their position, allow me a small exegetical detour. Much of the ink spilled on epistemic value focuses on the value of *knowledge*. By contrast, I am concerned with the value of *understanding*. Nevertheless, we can appropriate certain distinctions and frameworks from the former discussion. In particular, I recast Pritchard's

(2010) insightful mapping of the issues about knowledge's value in terms of understanding's value.

Pritchard asks, "Why is knowledge more valuable in kind and not simply in degree than that which falls short of knowledge?"[5] The corresponding question for understanding will be:

- *Distinctive Value Question:* Why are increases in understanding more valuable *in kind* than that which falls short of these increases?

Let's say that understanding is *distinctively valuable* if this question's central presupposition is true. Hence, I deny that understanding is distinctively valuable.

Having said this, I can see the appeal of wanting an answer to the Distinctive Value Question. Such an answer would make understanding special. Moreover, an answer to the Distinctive Value Question doubles as an answer to the Basic Value Question, but the converse does not hold. So, the Distinctive Value Question might then seem like a holier grail than the Basic Value Question.

Nevertheless, I think something is deeply amiss with the Distinctive Value Question. Consequently, I begin by discussing its underlying assumptions (Section 8.2.1). I then discuss why, if the EKS Model is true, understanding is not distinctively valuable (Section 8.2.2). Next, I highlight difficulties with other accounts of understanding's distinctive value (Section 8.2.3). I then argue that the motivations for even *asking* the Distinctive Value Question are mistaken (Section 8.2.4). I conclude by generalizing my objections, so as to dim the prospects of any positive answer to this question (Section 8.2.5).

8.2.1. Framing the Issue

Ultimately, I will argue for two theses. First, the Distinctive Value Question does not admit of an answer; second, it's not even worth asking. The first of these positions has garnered more attention – once again by Pritchard when discussing *knowledge*'s value. Specifically, Pritchard offers three different responses to knowledge's analogue to the Distinctive Value Question that I will repurpose for my discussion of understanding's value.

---

[5] Pritchard calls this the "tertiary value problem." He also discusses the "primary" and "secondary" value problems for knowledge. In effect, their analogues for understanding are answered in Section 8.1.

First, *validationists* accept the presuppositions of the Distinctive Value Question and endeavor to provide a direct answer to it. In other words, validationists cite certain features of understanding that make it more valuable in kind than that which falls short of it. It's fair to say that those who have discussed understanding's epistemic value are validationists. Indeed, validationism about understanding's value is often motivated by a less rosy outlook about knowledge's distinctive value (Kvanvig 2003, 2009b, 2009a; Pritchard 2008, 2009a, 2010).

However, in principle, a *revisionist* response to the Distinctive Value Question is also available. Revisionism has three stages. First, on this view, one rejects the central presupposition of the Distinctive Value Question. In other words, one denies that understanding is more valuable in kind than any proper subset of its parts. Second, one attempts to rescue the value-driven inquiry by identifying an epistemic status that is distinct from understanding, but nevertheless resembles it. Third, one argues that this alternative epistemic status is more valuable in kind than any proper subset of its parts. The result is a kind of error-theory: we *thought* that understanding was distinctively valuable, but in fact it was really this nearby epistemic status that is the genuine bearer of distinctive value. Essentially, Kvanvig and Pritchard are revisionists about *knowledge*'s distinctive value, which leads them to be validationists about *understanding*'s distinctive value.

The third and final response, *fatalism*, finds the revisionist too apologetic. Like revisionists, fatalists deny the Distinctive Value Question's central presupposition, but do not seek to resuscitate the value-driven inquiry by finding some surrogate epistemic status that is distinctively valuable. On this view, there is typically something fundamentally mistaken about the structure of the Distinctive Value Question, such that asking its analogue for any epistemic state – or at least any state bearing close resemblance to understanding – rests on a deep confusion.

I will be arguing for fatalism. Consequently, I owe you a diagnosis about why the Distinctive Value Question is misguided. In other words, I need to tell you why the very *motivations* that lead to the *asking* of the Distinctive Value Question rest on a mistake.

Philosophical discussions of these motivations are rare and brief. Nevertheless, Pritchard is once again our exemplar, providing two motivations for the Distinctive Value Question. First, suppose that understanding is more valuable than any proper subset of its parts, but that this difference in value is only one of degree. Then we will have answered the Basic but not the Distinctive Value Question. Discussing the analogous case with knowledge, Pritchard (2010, 7–8) highlights one loss we might thereby suffer:

The problem with this "continuum" account of the value of knowledge, however, is that it fails to explain why the long history of epistemological discussion has focused specifically on the stage in this continuum of value that knowledge marks rather than some other stage (such as a stage just before the one marked out by knowledge, or just after).

Of course, unlike knowledge, understanding does not enjoy a "long history of epistemological discussion." Nevertheless, a failure to answer the Distinctive Value Question might cast doubts about why we should have any *future* philosophical discussions about understanding. Call this the *Philosophical Focus Motivation*, since it recruits answers to the Distinctive Value Question to tell us why understanding is worthier of philosophical attention than similar epistemic standings.

A second motivation for pursuing the Distinctive Value Question is that we might consider understanding to be *finally* valuable (i.e., valuable for its own sake). If that which falls just short of understanding is only incrementally less valuable, then perhaps it's just as fitting to set this lesser status as our epistemic goal. If that's correct, then understanding is, in some sense, gratuitous or extravagant. As Pritchard (2010, 8) puts it:

> ... support for [this] problem comes from the fact that we often treat knowledge as being, unlike lesser epistemic standings, *precious*, in the sense that its value is not merely a function of practical import. But if that's correct, then knowledge must be the kind of thing that, unlike that which falls short of knowledge, is valuable for its own sake. That is, it must be non-instrumentally – i.e., *finally* – valuable.

Thus, transposing this idea to understanding, our axiological itch will only be scratched if understanding is precious in the sense of being finally valuable. Furthermore, if it is not distinctively valuable, then this final value is threatened. Call this the *Final Value Motivation*.

Hence, as a fatalist, my goals are threefold. First, I must show that an increase in understanding is not more valuable in kind than that which falls short of that increase. Second, I must defuse the Philosophical Focus Motivation by showing either that understanding is not worthy of philosophical attention, or that such attention is warranted even if the Distinctive Value Question cannot be answered positively. Third, I must do the same with the Final Value Motivation, showing that understanding is either not finally valuable or would be finally valuable even if the Distinctive Value Question cannot be answered positively.

Note that one way of extinguishing each of these motivations might be called *hyper-fatalist*. On such a view, understanding is neither worthy of philosophical focus nor finally valuable. As I hope the efforts of this book suggest, I am no hyper-fatalist. Rather, I favor a second, *therapeutic* fatalism, in which we get everything that we could have asked for – philosophical focus on a finally valuable kind of understanding – without having to answer a misguided question.

8.2.2.   *Fatalism and the EKS Model*

So, why should we view the Distinctive Value Question with such suspicion? After all, don't the arguments in Section 8.1.1 point toward the redemption of understanding's distinctive value? In a word: no.

To see this, three points are in order. First, keep in mind that anything that falls short of one of my four improvements in understanding but that has the same *kind* of value robs this improvement of its distinctive value. It also robs the resulting understanding of this distinctive value. Examples of this will be discussed later.

Second, observe that a scientific improvement entails the three other kinds of improvements. After all, a scientific improvement requires a true belief about a correct explanation (i.e., an explanatory improvement) and is the result of better scientific explanatory evaluation (which is the home of improved considerations and comparisons.) Hence, to disprove the distinctive value of understanding, we simply need to find a cognitive achievement that falls short of a scientific improvement. I will argue that there are just such achievements.

Third, recall our previous points about cognitive achievements. The basic structure of a cognitive achievement is a cognitive success (i.e., a true belief) that is primarily because of an ability, where that ability is significant or involves the overcoming of a significant obstacle. The cognitive success involved in scientific improvements is a safe belief in an explanation. The cognitive abilities involved in scientific improvements are the consideration and comparison of alternative explanations of the target phenomenon.

With these points in tow, understanding's distinctive value hits the skids. For example:

> Nate has extraordinary eyesight and lip-reading skills. From across a smoke-filled laboratory, he eavesdrops on a conversation between two scientists, Molly and Kate. Molly is explaining to Kate that the smoke is because of a mixture of potassium chlorate, charcoal, and anthracene. Furthermore, Molly knows this to be the case and is asserting this claim with sincerity. Nearly nobody else in the world

could have eavesdropped on this conversation, but Nate is just that good.

The relevant issue isn't whether Nate understands or not. Rather, the central issue is whether Nate has a cognitive achievement. On this front, the answer is yes: Nate clearly has a true belief that is the result of significant abilities (eyesight and lip-reading) and the overcoming of significant obstacles (a smoke-filled room). Moreover, this cognitive achievement falls short of a scientific improvement. Nate's belief is safe, but not because of his SEEing. Nevertheless, he deserves credit for his true belief.

Thus, Nate has a cognitive achievement that requires less than a scientific improvement. Nevertheless, his achievement exhibits the same kinds of epistemic value that are found in a scientific improvement. He has true beliefs about the correct explanation of the smoke and hence enjoys the same value as explanatory improvements. He has earned those beliefs through significant cognitive ability and hence has the "achievement-value" characteristic of scientific improvements. He has exercised abilities that have instrumental epistemic value, just as one gets with improved considerations and comparisons.

8.2.3. Validationism Invalidated

All else being equal, a view that does not respect the intuition that understanding plays a special role in our epistemic undertakings is worse for wear than one that does. So, one might take the preceding as a strike against the EKS Model. Of course, any challenger to the EKS Model would have to spot faults in the reasoning of the previous chapters, and, as any aficionado of the Preface Paradox knows, I am bound to have erred somewhere. Putting this disclaimer aside, I will consider some other arguments about understanding's distinctive value.

One prominent idea is to conceive of understanding as a cognitive achievement in broader ways than the EKS Model does (Carter and Gordon 2014; Pritchard 2010). Note that this immediately blunts the force of the previous section. Scientific improvements are a fairly specific kind of cognitive achievement, and this is one reason that it is easy to find cognitive achievements that fall short of them. So, if understanding is a far broader kind of cognitive achievement, then the set of counterexamples of this kind contracts significantly. Essentially, this would recast the Scientific Achievement Argument as follows:

General Achievement Argument

GA1.   Achievements are successes that are primarily because of ability, where the success in question either involves the overcoming of a significant obstacle or the exercise of a significant level of ability.

GA2.   Understanding is a cognitive success that is primarily because of cognitive ability, where the success in question either involves the overcoming of a significant obstacle or the exercise of a significant level of ability.

GA3.   Achievements have final value.

GA4.   ∴ Understanding has final value (GA1–GA3).

Because GA1 and GA3 in this argument are the same as SA1 and SA3 in the Scientific Achievement Argument, my only misgiving is with GA2. As I will show, it is either false or it trivializes the enterprise of answering the Distinctive Value Question.

First, consider cases in which GA2 is false. Some instances of low-grade understanding are not strong achievements in the sense that GA2 requires. For instance, I look at a wall and instantly understand why it is white – because it was painted white, based on the rather unremarkable inductive evidence that most white walls have been painted white. This involves neither the overcoming of a significant obstacle nor the exercising of a significant skill.

Carter and Gordon (2014) have highlighted this shortcoming in Pritchard's version of the General Achievement Argument. They attempt to remedy it by revising the General Achievement Argument as follows:

GA2*.   *Rich objectual* understanding is a cognitive success that is primarily because of cognitive ability, where the success in question either involves the overcoming of a significant obstacle or the exercise of a significant level of ability.

Here, "rich objectual understanding" is most readily associated with mastery of a subject matter. As we saw in Chapter 4, this kind of objectual understanding is reducible to "rich" explanatory understanding. However, here lies the rub: construed as a kind of *outright* understanding, rich explanatory understanding is altogether arbitrary: it's simply naming an arbitrary point on the continuum of understanding that too easily happens to be good enough to be a cognitive achievement. The vapidity of this maneuver becomes obvious when you realize that you can apply it to many other epistemic standings without a dint of argumentation: simply identify the kind of knowledge, justified true belief, insight, or the like that involves cognitive abilities, then call it "rich knowledge," "rich justified true belief," or "rich insight," and you've "answered" the Distinctive Value Question. A qualitative difference in understanding's value *appears* to rest on a qualitative difference in its nature. Yet everything here appears to be either a mere difference in degree or a merely incidental difference in kind.

Indeed, precisely because of this, even if one doesn't accept my quasi-explanationist stance from Chapter 4, this worry persists. As Carter and Gordon admit, some objectual understanding is not rich. We can understand a subject matter very superficially, as is my understanding of paints and walls. However, this means that rich objectual understanding is no less arbitrary than rich explanatory understanding. Thus, quasi-explanationism or not, simply adding the word "rich" to one's favorite kind of understanding is little more than an incantation that answers the Distinctive Value Question a bit too conveniently.

Here is the deeper lesson: failure to respect the fact that understanding admits of degrees leads to a seductive but ill-fated strategy. The strategy goes something like this: we see that some epistemic state that we valued – say knowledge – faces steep challenges with respect to a Value Question. We survey the neighborhood of nearby epistemic states, and we find that some instances of understanding seem to avoid those liabilities. That's the source of the seduction. However, here is how that strategy can quickly become ill-fated: since understanding comes in degrees, one must avoid cherry-picking an arbitrary point on its spectrum and then giving it a special label after the fact.

8.2.4.   No Motivation

Thus far, I have argued that if the EKS Model is correct, then understanding is not distinctively valuable. I have also shown that other attempts to validate the Distinctive Value Question rest on *ad hoc* maneuvers that pay insufficient attention to how understanding admits of degrees. Furthermore, these latter attempts rest on accounts of the nature of understanding that I have critiqued in previous chapters.[6] If these arguments are sound, then the very presuppositions of the Distinctive Value Question are false, and so endeavoring an answer is, as the ancients said, like milking the

---

[6] My critique of Pritchard is in Chapters 3 and 7; my critique of Carter and Gordon, in Chapter 4.

he-goat, leaving those who await that answer with the unenviable task of holding the sieve.

However, I do not want to leave things there. I also promised to offer a diagnostic account of why the very motivations for asking the Distinctive Value Question were unwarranted. Recall that the first was what I dubbed the Philosophical Focus Motivation: if understanding is worthy of philosophical attention, then it must be distinctively valuable. I noted that one might go hyper-fatalist and deny that understanding is worthy of philosophical attention. Alternatively, I will offer a therapeutic fatalism in which understanding is worthy of attention even if it is not distinctively valuable.

Here, I can be brief: scientific improvements are increases in understanding, and scientific improvements are worthy of philosophical attention. Moreover, one reason they deserve this focus is that they are significant cognitive achievements. So far as I can tell, Pritchard, Carter, and Gordon can agree with all of this, even if their enthusiasm for science is less effusive than my own. The point is that none of this hinges on scientific improvements being distinctively valuable.

Furthermore, abandoning the search for distinctively valuable kinds of understanding allows us to focus on a more interesting array of epistemic statuses. For instance, consider our three other improvements (explanatory improvements, improved consideration, and improved comparison). These are even less promising ways of establishing understanding's distinctive value than scientific improvements. Nevertheless, we have already seen that they deserve our philosophical attention. For instance, recall that my discussion of idealizations involved a kind of acceptance that is effective in meeting various context-specific scientific goals (see Chapter 6). For all that I've argued, these goals could be achieved without the idealizations, and hence their value is not distinctive. Nevertheless, idealizations raise very interesting questions about understanding's relationship to truth. Consequently, we should not restrict our philosophical focus to distinctively valuable epistemic statuses.

More generally, the *kind* of value that a component of understanding has is orthogonal to its philosophical relevance. What matters is the *degree*. If something were distinctively valuable but barely better than worthless, it surely would not warrant philosophical focus. Rather, it seems as if philosophy should fix its gaze on highly valuable, things or on those things that raise puzzles about highly valuable things. Neither of these foci needs to be distinctively valuable.

The Final Value Motivation also purported to underscore the Distinctive Value Question's importance. The general idea here is that understanding can only have final value if it also has distinctive value. Here is one place where degrees of understanding matter quite a bit. They also belie this motivation.

Suppose that you are seeking to understand why something is the case. You can either have pretty good scientific knowledge of one correct explanation of this phenomenon, or you can have even better scientific knowledge of most of the correct explanations of that phenomenon, as well as their interrelations. Furthermore, suppose that the scientific knowledge gained in the second option entails the knowledge gained in the first option. Which would it be fitting to favor for its own sake? Clearly the second, which suggests that having better understanding has final value. However, there is no reason that the second option must have a kind of value that the first option lacks. Hence, understanding that is finally valuable need not be distinctively valuable. Note that even if one wants to claim that both statuses have final value, it's still the case that the more demanding kind of understanding is not distinctively valuable. One needs to show that there is something special about aiming for less understanding.[7]

8.2.5.   Shoring Up the Fatalist Foothold

Thus, we have seen that the very reasons for *asking* the Distinctive Value Question are groundless. If this is correct, then fatalism is the appropriate response to this question. We have offered three reasons for this verdict. First, the presupposition of the Distinctive Value Question is false: understanding is not more valuable in kind than that which falls short of it. Second, philosophers have been well served by focusing on understanding that is not distinctively valuable. Third, there appear to be finally valuable instances of understanding that are not distinctively valuable.

However, perhaps I have only shown that validationism about understanding's distinctive value should not appeal to some kind of "Achievement Argument." While I can't claim to have eliminated this possibility, I hope to push the dialectic further in the fatalist direction. To that end, I will show that my arguments in this section generalize quite widely. I will also indicate some of the ways in which non-fatalists could reply to my arguments, primarily to show how much work needs to be done to revive the Distinctive Value Question. If this is right, then fatalism

---

[7] Thanks to Emily Sullivan for raising this point.

should enjoy the default position in subsequent discussions of the Distinctive Value Question.

I will offer three reasons for this stance. First, outright and better understanding lack the kind of invariant features that are liable to have a tight conceptual link with anything distinctively valuable. Some instances of understanding will have this putatively distinctive value; other instances will not. Hence, one is left with two options: gerrymander a concept such as "rich understanding" to accommodate one's favorite thesis about epistemic value or look to ideal understanding as the only non-arbitrary, distinctively valuable kind of understanding. However, both options face some stiff challenges. We've already seen the problem with the first option in discussing Carter and Gordon's work. "Rich understanding" can't simply be a post hoc label for "the kind of understanding that would answer the Distinctive Value Question." That would be little more than verbal subterfuge.

What, then, of the second option; i.e., of leaning on *ideal* understanding as the bearer of distinctive value? This faces the problem of undercutting the Philosophical Focus Motivation. Recall that an answer to the Distinctive Value Question is supposed to justify why understanding is worthy of philosophical attention. However, if only *ideal* understanding is tightly linked with whatever is distinctively valuable, this motivation loses much of its force. Ideal understanding is utopian: no human actually has it. The philosophical study of *non-ideal* understanding – the understanding that flesh-and-blood humans are capable of attaining – seems a far worthier pursuit than ideal understanding. After all, we mortals only trade in non-ideal understanding.

Compare this with the following: being able to perceive ideally is almost certainly of distinctive value, yet few would argue that perception is worthy of philosophical investigation because of this ideal. On the contrary: perception is worthy of philosophical investigation because it is valuable to so much of our *everyday* knowledge and so much of our *everyday* mental life. Note that neither horn makes any special appeal to achievements. Hence these are very general problems.

Additionally, my critique of the Final Value Motivation undermines a wide class of arguments of which the Scientific and General Achievement Arguments are just instances. Essentially, these arguments all have the following form:

1. Epistemic status E is F.
2. F's have final value.

3. All finally valuable statuses are distinctively valuable.
4. ∴ *E* is distinctively valuable (1–3).

This will be validationist if our epistemic status *E* is some kind of understanding; revisionist, otherwise. The Final Value Motivation claims that we need to answer the Distinctive Value Question in order to account for *E*'s final value. However, as we saw earlier, the third premise is false. Moreover, as this argument pattern makes clear, once you've established the first two premises, you've accounted for *E*'s final value. Hence, distinctive value is an idle wheel. Importantly, this general structure does not require that *F* be about achievements.

Third, any straightforward attempt to rescue the Distinctive Value Question (either through validationist or revisionist means) faces a difficult tradeoff in linking distinctive and final value. One way to appreciate this point is to recast the Final Value Motivation as trying to *explain* how a certain property makes understanding finally valuable. On such a view, we might think (very loosely) of distinctive value as some kind of intermediate mechanism between an epistemic status and its final value. In this case, a sound instance of the following will restore this revised version of the Final Value Motivation:

1. Epistemic status *E* is *F*.
2. *F*'s have distinctive value.
3. All distinctively valuable statuses are finally valuable.
4. ∴ *E* is finally valuable (1–3).

However, here is where the tradeoff rears its ugly head. On the one hand, a distinctively valuable status will generally exhibit the minimum of a certain kind of value. On the other hand, a finally valuable status will pull us away from this minimum. For instance, this argument will still not work for achievements, for we have already seen that some achievements are not distinctively valuable (e.g., ideal understanding). So, the second premise will be false in this instance. Essentially, this means that the only distinctively valuable achievements are those that are just barely achievements, since anything that falls short of them will miss out on the distinctive value of achievements. By contrast, more impressive achievements will be more valuable than more modest achievements only as a matter of degree, and hence won't be distinctively valuable.

However, minimal achievements are not the ones that seem to be finally valuable. After all, it is fitting to favor getting most of the credit for a success rather than the bare minimum. Hence, it appears that the more suitable an achievement is for being distinctively valuable, the less suitable it is for being finally valuable. So far as I can tell, this is a general feature of this argument and works when we turn our attention to other potential candidates for this intermediate mechanism. For instance, consider the value of accuracy: having something that is barely accurate is distinctively valuable, but we should strive to be more than barely accurate.

Admittedly, there are dilemmas that might be surmountable and bullets that might be bitten in the preceding. One *might* find a way of redeeming "rich understanding" as a non-arbitrary kind of understanding. One *might* also note that there is a deep link between ideal and outright understanding on my account, such that the former's value *somehow* informs the latter's. One *might* also offer an argument that has a different structure than those critiqued earlier. One *might* look for different motivations for the Distinctive Value Question. One *might* adopt a revisionist stance and show that some adjacent epistemic status avoids these worries. Here's the kicker: all of these are non-trivial and haven't been done. Consequently, the burden of proof falls squarely on the non-fatalist's shoulders.

Indeed, to treat these possibilities as anything other than licenses for a hazy optimism underestimates the prominence of the General Achievement Argument in establishing understanding's distinctive value. Even more than its counterpart in discussions of knowledge's value, the General Achievement Argument has been the validationists' weapon of choice. So far as I can tell, this is Pritchard's *only* argument for the distinctive value of understanding – ditto for Carter and Gordon (2014).[8]

Finally, let's return to the General Value Question with which we began:

• Do understanding and knowledge have the same kinds of epistemic value?

We have seen that understanding's value is only greater than that which falls short of it. Pritchard also grants that we can give a very similar answer to the corresponding question about knowledge. Consequently, much like ability, truth, luck, and objectual structure, there is nothing about epistemic value that distinguishes understanding from its neighboring concepts.

---

[8] While Kvanvig (2003, 2009a, 2009b) is congenial to the General Achievement Argument, he offers another source of understanding's value: subjective justification. Since I have navigated all of the major problems concerning the nature of understanding without having to lean on subjective justification, a further argument would be needed to show that understanding requires subjective justification. Here, I sympathize with Grimm (2016), who offers some general arguments against any strong internalist requirement on understanding.

8.3. Conclusion: Understanding, Philosophy, and Scientific Knowledge

I have argued that understanding is valuable in some regards and not valuable in others. The good news is that understanding is more valuable than that which falls short of it. In particular, when science improves along its usual route of securing safer beliefs through careful consideration and comparison of plausible explanations, we have latched onto a cognitive achievement that exhibits a fundamental epistemic value – true belief in an explanation. That achievement also exhibits the instrumental epistemic value characteristic of our best scientific methods. Finally, as an achievement, scientific improvement is valuable in its own right.

Some will find this inadequate: understanding must not only be valuable in this way, but it must be more valuable in *kind* than that which falls short of it. I have argued that we should disabuse ourselves of this philosophical yearning. First, it is a yearning for something far out of reach – if not unattainable. Second, the motivations for raising questions about this distinctive value of understanding can be satisfied without jumping down the rabbit hole it invites.

In many ways, this "fatalism" about understanding's distinctive value echoes the ethos of my intellectual heroes – those philosophers of science who treated understanding as an afterthought and who are often associated with the received view of understanding. If understanding isn't distinctively valuable, there is a sense in which it is not special. The arguments of this book have consistently concluded in this fatalist or deflationary key. To make sense of understanding, we do not need special abilities beyond those that scientists deploy in confirming an explanation. We do not need some special, irreducible kind of objectual understanding. We do not need to agonize over alleged cases of understanding without explanation. If understanding makes departures from the truth, these departures only seem fancy because epistemologists have neglected scientific practice. Nor do we need understanding to have a special tolerance for luck. And we do not need to puzzle over a chimerical kind of epistemic value.

So, in a sense, I have inherited this wider-ranging fatalism. But my position can also be seen in a more positive light. If we take that position seriously, we have freed ourselves from the encumbrances of philosophical theorizing gone too far. It is high time to expand our vistas, now liberated from that theorizing's strictures. We philosophers should not saddle ourselves with high-minded epistemologies, axiologies, or some misguided search for a "special" ingredient that adds an empty veneer of depth to our theory of understanding. We should find comfort in embracing the wisdom of scientific practice where we can. In doing this, we will find that there is more than enough value and understanding if our philosophical focus remains fixed on the march of scientific progress.

Bibliography

Achinstein, Peter. 1983. *The nature of explanation.* New York: Oxford University Press.

Bartelborth, Thomas. 2002. "Explanatory unification." *Synthese* 130 (1):91–107.

Batterman, Robert W. 2000. "A 'modern' (= Victorian?) attitude towards scientific understanding." *Monist* 83 (2):228–257.

Batterman, Robert W. 2002. *The devil in the details: asymptotic reasoning in explanation, reduction and emergence.* Oxford: Oxford University Press.

Batterman, Robert W., and Collin C. Rice. 2014. "Minimal model explanations." *Philosophy of science* 81 (3):349–376. doi: 10.1086/676677.

Bjorken, James. 1967. "Theoretical ideas on inelastic electron and muon scattering." In *1967 international symposium on electron and photon interactions at high energies,* edited by S. M. Berman, 109–127. Stanford: International Union of Pure and Applied Physics, US Atomic Energy Commission.

Bogen, Jim. 2005. "Regularities and causality: generalizations and causal explanations." *Studies in history and philosophy of biological and biomedical sciences* 36 (2):397–420.

Bokulich, Alisa. 2008. "Can classical structures explain quantum phenomena?" *The British journal for the philosophy of science* 59 (2):217–235. doi: 10.1093/bjps/axn004.

Bonjour, Laurence. 1998. *In defense of pure reason: a rationalist account of a priori justification.* Cambridge: Cambridge University Press.

Brandom, Robert. 1994. *Making it explicit: reasoning, representing, and discursive commitment.* Cambridge, MA: Harvard University Press.

Braverman, Mike, John Clevenger, Ian Harmon, Andrew Higgins, Zachary Horne, Joseph Spino, and Jonathan Waskan. 2012. "Intelligibility is necessary for scientific explanation, but accuracy may not be." In *Proceedings of the thirty-fourth annual conference of the cognitive science society,* edited by Naomi Miyake, David Peebles, and Richard Cooper. Cognitive Science Society.

Brogaard, Berit. Manuscript. "I know. Therefore, I understand." http://www.academia.edu/download/3458713/BrogaardOnKvanvig2.pdf

Cao, Tian Yu. 2010. *From current algebra to quantum chromodynamics: a case for structural realism.* Cambridge: Cambridge University Press.

Carter, J. Adam, and Emma C. Gordon. 2014. "Objectual understanding and the value problem." *American philosophical quarterly* 51 (1):1–13.

Cartwright, Nancy. 1983. *How the laws of physics lie.* Oxford: Clarendon Press.

Cartwright, Nancy. 2004. "From causation to explanation and back." In *The future for philosophy*, edited by Brian Leiter, 230–245. Oxford: Oxford University Press.

Chang, Hasok. 2012. *Is water H₂O?: evidence, realism and pluralism, Boston studies in the philosophy of science.* Dordrecht: Springer Verlag.

Chisholm, Roderick M. 1966. *Theory of knowledge, Foundations of philosophy series.* Englewood Cliffs: Prentice-Hall.

Code, Lorraine. 1987. *Epistemic responsibility.* Hanover, NH: Published for Brown University Press by University Press of New England.

Cohen, L. Jonathan. 1992. *An essay on belief and acceptance.* Oxford: Clarendon Press.

Craver, Carl F. 2007. *Explaining the brain: mechanisms and the mosaic unity of neuroscience.* Oxford: Clarendon Press.

Cushing, James T. 1991. "Quantum theory and explanatory discourse: endgame for understanding?" *Philosophy of science* 58 (3):337–358.

David, Marian. 2001. "Truth as the epistemic goal." In *Knowledge, truth, and duty: essays on epistemic justification, responsibility, and virtue*, edited by Matthias Steup, 151–169. New York: Oxford University Press.

De Regt, Henk W. 2004. "Discussion note: making sense of understanding." *Philosophy of science* 71:98–109.

De Regt, Henk W. 2009a. "The epistemic value of understanding." *Philosophy of science* 76 (5):585–597. doi: 10.1086/605795.

De Regt, Henk W. 2009b. "Understanding and scientific explanation." In *Scientific understanding: philosophical perspectives*, edited by Henk W. De Regt, Sabina Leonelli, and Kai Eigner, 21–42. Pittsburgh: University of Pittsburgh Press.

De Regt, Henk W. 2015. "Scientific understanding: truth or dare?" *Synthese* 192 (12):3781–3797. doi: 10.1007/s11229-014-0538-7.

De Regt, Henk W., and Dennis Dieks. 2005. "A contextual approach to scientific understanding." *Synthese* 144 (1):137–170.

DePaul, Michael R., and Stephen R. Grimm. 2007. "Review essay on Jonathan Kvanvig's *The value of knowledge and the pursuit of understanding*." *Philosophy and phenomenological research* 74 (2):498–514.

Devitt, Michael. 2014. "There is no a priori." In *Contemporary debates in epistemology*, edited by Matthias Steup, John Turri, and Ernest Sosa, 185–194. Malden, MA: Wiley Blackwell.

Diéguez, Antonio. 2013. "When do models provide genuine understanding, and why does it matter?" *History and philosophy of the life sciences* 35 (4):599–620.

Diéguez, Antonio. 2015. "Scientific understanding and the explanatory use of false models." In *The future of scientific practice: 'Bio-TechnoLogos'*, edited by Marta Bertolaso, 161–178. London: Pickering & Chatto.

Díez, José, Kareem Khalifa, and Bert Leuridan. 2013. "General theories of explanation: buyer beware." *Synthese* 190 (3):379–396. doi: 10.1007/s11229-011-0020-8.

Douglas, Heather E. 2009. "Reintroducing prediction to explanation." *Philosophy of science* 76 (4):444–463.

Doyle, Yannick, Spencer Egan, Noah Graham, and Kareem Khalifa. Manuscript. "Non-factive understanding: a statement and defense." Available from the author.

Elgin, Catherine Z. 2004. "True enough." *Philosophical issues* 14 (1):113–131. doi: 10.1111/j.1533-6077.2004.00023.x.

Elgin, Catherine Z. 2007. "Understanding and the facts." *Philosophical studies* 132 (1):33–42.

Elgin, Catherine Z. 2009a. "Exemplification, idealization, and scientific understanding." In *Fictions in science: philosophical essays on modeling and idealization*, edited by Mauricio Suárez, 77–91. London: Routledge.

Elgin, Catherine Z. 2009b. "Is understanding factive?" In *Epistemic value*, edited by Adrian Haddock, Allan Millar, and Duncan Pritchard, 322–330. Oxford: Oxford University Press.

Faye, Jan. 2007. "The pragmatic-rhetorical theory of explanation." In *Rethinking explanation*, edited by Johannes Persson and Petri Ylikoski, 43–68. Dordrecht: Springer.

Friedman, Michael. 1974. "Explanation and scientific understanding." *Journal of philosophy* 71 (1):5–19.

Galilei, Galileo. 1914. *Dialogues concerning two new sciences*. Translated by Henry Crew and Antonio de Salvio. New York: Macmillan. Original edition, 1638.

Garfinkel, Alan. 1981. *Forms of explanation: rethinking the questions in social theory*. New Haven: Yale University Press.

Gendler, Tamar Szabó. 1998. "Galileo and the indispensability of scientific thought experiment." *The British journal for the philosophy of science* 49 (3):397–424. doi: 10.1093/bjps/49.3.397.

Gijsbers, Victor. 2013. "Understanding, explanation, and unification." *Studies In history and philosophy of science part A* 44 (3):516–522. doi: 10.1016/j.shpsa.2012.12.003.

Gijsbers, Victor. 2014. "Unification as a measure of natural classification." *Theoria: revista de teoría, historia y fundamentos de la ciencia* 29 (1):71–82.

Glymour, Bruce. 2007. "In defence of explanatory deductivism." In *Causation and explanation*, edited by Joseph Keim Campbell, Michael O'Rourke, and Harry S. Silverstein, 133–154. Cambridge, MA: MIT Press.

Goldman, Alvin I. 1976. "Discrimination and perceptual knowledge." *Journal of philosophy* 73 (20):771–791.

Gordon, Emma C. 2012. "Is there propositional understanding?" *Logos & episteme* 3 (2):181–192.

Greco, John. 2009. *Achieving knowledge*. Cambridge: Cambridge University Press.

Greco, John. 2013. "Episteme: knowledge and understanding." In *Virtues and their vices*, edited by Kevin Timpe and Craig A. Boyd, 285–301. Oxford: Oxford University Press.

Grimm, Stephen R. 2006. "Is understanding a species of knowledge?" *The British journal for the philosophy of science* 57 (3):515–535.

Grimm, Stephen R. 2009a. "Epistemic normativity." In *Epistemic value*, edited by Adrian Haddock, Allan Millar, and Duncan Pritchard, 243–264. Oxford: Oxford University Press.

Grimm, Stephen R. 2009b. "Reliability and the sense of understanding." In *Scientific understanding: philosophical perspectives*, edited by Henk W. De Regt, Sabina Leonelli, and Kai Eigner, 83–99. Pittsburgh: Univeristy of Pittsburgh Press.

Grimm, Stephen R. 2010. "The goal of understanding." *Studies in the history and philosophy of science* 41 (4):337–344. doi: 10.1016/j.shpsa.2010.10.006.

Grimm, Stephen R. 2014. "Understanding as knowledge of causes." In *Virtue epistemology naturalized*, edited by Abrol Fairweather, 329–345. Dordecht: Springer International Publishing.

Grimm, Stephen R. 2016. "Understanding and transparency." In *Explaining understanding: new essays in epistemology and the philosophy of science*, edited by Stephen R. Grimm, Christoph Baumberger and Sabine Ammon, 212–229. New York: Routledge.

Hacking, Ian. 1982. "Experimentation and scientific realism." *Philosophical topics* 13:71–88.

Harman, G. 1986. *Change in view: principles of reasoning*. Cambridge: MIT Press.

Hazen, T. C., E. A. Dubinsky, T. Z. DeSantis, G. L. Andersen, Y. M. Piceno, N. Singh, J. K. Jansson, A. Probst, S. E. Borglin, J. L. Fortney, W. T. Stringfellow, M. Bill, M. E. Conrad, L. M. Tom, K. L. Chavarria, T. R. Alusi, R. Lamendella, D. C. Joyner, C. Spier, J. Baelum, M. Auer, M. L. Zemla, R. Chakraborty, E. L. Sonnenthal, P. D'Haeseleer, H. Y. N. Holman, S. Osman, Z. Lu, J. D. Van Nostrand, Y. Deng, J. Zhou, and O. U. Mason. 2010. "Deep-sea oil plume enriches indigenous oil-degrading bacteria." *Science* 330 (6001):204–208. doi: 10.1126/science.1195979.

Hempel, Carl G. 1965. *Aspects of scientific explanation, and other essays in the philosophy of science*. New York: Free Press.

Hempel, Carl G. 1966. *Philosophy of natural science, Prentice-Hall foundations of philosophy series*. Englewood Cliffs: Prentice-Hall.

Hills, Alison. 2009. "Moral testimony and moral epistemology." *Ethics* 120 (1):94–127. doi: 10.1086/648610.

Hills, Alison. 2015. "Understanding why." *Noûs* 49 (2):661–688. doi: 10.1111/nous.12092.

Hindriks, Frank. 2013. "Explanation, understanding, and unrealistic models." *Studies in history and philosophy of science part A* 44 (3):523–531. doi: 10.1016/j.shpsa.2012.12.004.

Hitchcock, Christopher R. 1999. "Contrastive explanation and the demons of determinism." *The British journal for the philosophy of science* 50 (4):585–612.

Hitchcock, Christopher R., and James Woodward. 2003. "Explanatory generalizations, part II: plumbing explanatory depth." *Noûs* 37 (2):181–199.

Humphreys, Paul. 1989. *The chances of explanation*. Princeton: Princeton University Press.

Jeffrey, Richard C. 1969. "Statistical explanation vs. statistical inference." In *Essays in honor of Carl G. Hempel*, edited by Nicholas Rescher, 104–113. Dordrecht: Springer Netherlands.

Kelp, Christoph. 2014. "Knowledge, understanding and virtue." In *Virtue epistemology naturalized*, edited by Abrol Fairweather, 347–360. Dordrecht: Springer International Publishing.

Kelp, Christoph. 2015. "Understanding phenomena." *Synthese* 192 (12):3799–3816. doi: 10.1007/s11229-014-0616-x.

Khalifa, Kareem. 2011. "Understanding, knowledge, and scientific antirealism." *Grazer philosophische studien* 83 (1):93–112.

Khalifa, Kareem. 2012. "Inaugurating understanding or repackaging explanation?" *Philosophy of science* 79 (1):15–37.

Khalifa, Kareem. 2013. "Understanding, grasping, and luck." *Episteme* 10 (1):1–17.

Khalifa, Kareem. 2015. "EMU defended: reply to Newman (2014)." *European journal for philosophy of science* 5 (3):377–385. doi: 10.1007/s13194-015-0112-2.

Khalifa, Kareem. 2016. "Must understanding be coherent?" In *Explaining understanding: new perspectives from epistemology and philosophy of science*, edited by Stephen R. Grimm, Christoph Baumberger, and Sabine Ammon, 139–164. London: Routledge.

Khalifa, Kareem, and Michael C. Gadomski. 2013. "Understanding as explanatory knowledge: the case of Bjorken scaling." *Studies in history and philosophy of science part A* 44 (3):384–392. doi: 10.1016/j.shpsa.2013.07.001.

Khalifa, Kareem, Jared Millson, and Mark Risjord. Forthcoming. "IBE: fundamentalism's failures." In *Best explanations: new essays on inference to the best explanation*, edited by Kevin McCain and Ted Poston. Oxford: Oxford University Press.

Kim, Jaegwon. 1994. "Explanatory knowledge and metaphysical dependence." *Philosophical issues* 5:51–69. doi: 10.2307/1522873.

Kitcher, Philip. 1989. "Explanatory unification and the causal structure of the world." In *Scientific explanation*, edited by Philip Kitcher and Wesley C. Salmon, 410–506. Minneapolis: University of Minnesota Press.

Kvanvig, Jonathan L. 2003. *The value of knowledge and the pursuit of understanding.* Cambridge: Cambridge University Press.

Kvanvig, Jonathan L. 2009a. "Responses to critics." In *Epistemic value*, edited by Adrian Haddock, Allan Millar, and Duncan Pritchard, 339–352. Oxford: Oxford University Press.

Kvanvig, Jonathan L. 2009b. "The value of understanding." In *Epistemic value*, edited by Adrian Haddock, Allan Millar, and Duncan Pritchard, 95–111. Oxford: Oxford University Press.

Lackey, Jennifer. 2007. "Why we don't deserve credit for everything we know." *Synthese* 158 (3):345–361.

Laudan, Larry. 1981. "A confutation of convergent realism." *Philosophy of science* 48 (1):19–49.

Lewis, David K. 1986. "Causal explanation." In *Philosophical papers*, 214–240. Oxford: Oxford University Press.

Lipton, Peter. 2004. *Inference to the best explanation.* 2nd ed. New York: Routledge. Original edition, 1991.

Lipton, Peter. 2009. "Understanding without explanation." In *Scientific understanding: philosophical perspectives,* edited by Henk W. De Regt, Sabina Leonelli, and Kai Eigner, 43–63. Pittsburgh: University of Pittsburgh Press.

Lycan, William G. 1988. *Judgement and justification.* Cambridge: Cambridge University Press.

Lycan, William G. 2002. "Explanation and epistemology." In *The Oxford handbook of epistemology,* edited by Paul K. Moser, 408–433. Oxford: Oxford University Press.

Lynch, Michael P. 2004. *True to life: why truth matters.* Cambridge, MA: MIT Press.

Lynch, Michael P. 2009. "The values of truth and the truth of values." In *Epistemic value,* edited by Adrian Haddock, Allan Millar, and Duncan Pritchard, 225–242. Oxford: Oxford University Press.

Machamer, Peter, Lindley Darden, and Carl F. Craver. 2000. "Thinking about mechanisms." *Philosophy of science* 67:1–25.

Marshall, Barry J., and J. Robin Warren. 1984. "Unidentified curved bacilli in the stomach of patients with gastritis and peptic ulceration." *The Lancet* 323 (8390):1311–1315.

Marshall, Barry J., Leah J. Barrett, Chandra Prakash, Richard W. McCallum, and Richard L. Guerrant. 1990. "Urea protects Helicobacter (Campylobacter) pylori from the bactericidal effect of acid." *Gastroenterology* 99 (3):697–702.

Marshall, Barry J., C. Stewart Goodwin, J. Robin Warren, Raymond Murray, Elizabeth D. Blincow, Stephen J. Blackbourn, Michael Phillips, Thomas E. Waters, and Christopher R. Sanderson. 1988. "A prospective double-blind trial of duodenal ulcer relapse after eradication of Campylobacter pylori." *The Lancet* 2:1437–1442.

Mitchell, Sandra D. 1997. "Pragmatic laws." *Philosophy of science* 64:468–479.

Mitchell, Sandra D. 2000. "Dimensions of scientific law." *Philosophy of science* 67 (2):242–265.

Mizrahi, Moti. 2012. "Idealizations and scientific understanding." *Philosophical studies* 160 (2):237–252. doi: 10.1007/s11098-011-9716-3.

Morris, Kevin. 2012. "A defense of lucky understanding." *The British journal for the philosophy of science* 63 (2):357–371. doi: 10.1093/bjps/axr023.

Newman, Mark. 2012. "An inferential model of scientific understanding." *International studies in the philosophy of science* 26 (1):1–26. doi: 10.1080/02698595.2012.653118.

Newman, Mark. 2013. "Refining the inferential model of scientific understanding." *International studies in the philosophy of science* 27 (2):173–197. doi: 10.1080/02698595.2013.813253.

Newman, Mark. 2014. "EMU and inference: what the explanatory model of scientific understanding ignores." *European journal for philosophy of science* 4 (1):55–74. doi: 10.1007/s13194-013-0075-0.

Norton, John D. 2007. "Causation as folk science." In *Causation, physics, and the constitution of reality: Russell's republic revisited*, edited by Huw Price and Richard Corry, 11–44. Oxford: Oxford University Press.

Nozick, Robert. 1981. *Philosophical explanations*. Cambridge, MA: Harvard University Press.

Orozco, Joshue. 2011. "Epistemic luck." *Philosophy compass* 6 (1):11–21. doi: 10.1111/j.1747-9991.2010.00365.x.

Peacocke, Christopher. 2008. *Truly understood*. Oxford: Oxford University Press.

Pickering, Andrew. 1984. *Constructing quarks: a sociological history of particle physics*. Chicago: University of Chicago Press.

Polanyi, Michael. 2009. *The tacit dimension*. Chicago: University of Chicago Press. Original edition, 1966.

Popper, Karl Raimund. 1963. *Conjectures and refutations: the growth of scientific knowledge*. London: Routledge & Kegan Paul.

Pritchard, Duncan. 2005. *Epistemic luck*. Oxford: Oxford University Press.

Pritchard, Duncan. 2007. "Recent work on epistemic value." *American philosophical quarterly* 44 (2):85.

Pritchard, Duncan. 2008. "Knowing the answer, understanding, and epistemic value." *Grazer philosophische studien* 77:325–339.

Pritchard, Duncan. 2009a. "Knowledge, understanding and epistemic value." *Royal institute of philosophy supplements* 64:19–43. doi: 10.1017/S1358246109000046.

Pritchard, Duncan. 2009b. "Safety-based epistemology: whither now?" *Journal of philosophical research* 34:33–45.

Pritchard, Duncan. 2010. "Knowledge and understanding." In *The nature and value of knowledge: three investigations*, edited by Duncan Pritchard, Allan Millar, and Adrian Haddock, 3–90. Oxford: Oxford University Press.

Pritchard, Duncan. 2012. "Anti-luck virtue epistemology." *Journal of philosophy* 109 (3):247–279.

Pritchard, Duncan. 2014. "Knowledge and understanding." In *Virtue epistemology naturalized*, edited by Abrol Fairweather, 315–327. Dordrecht: Springer International Publishing.

Railton, Peter. 1978. "A deductive-nomological model of probabilistic explanation." *Philosophy of science* 45 (2):206–226.

Railton, Peter. 1981. "Probability, explanation, and information." *Synthese* 48:233–256.

Riaz, Amber. 2015. "Moral understanding and knowledge." *Philosophical studies* 172 (1):113–128. doi: 10.1007/s11098-014-0328-6.

Riggs, Wayne. 2009. "Understanding, knowledge, and the meno requirement." In *Epistemic value*, edited by Adrian Haddock, Alan Millar, and Duncan Pritchard, 331–338. Oxford: Oxford University Press.

Riordan, Michael. 1987. *The hunting of the quark: a true story of modern physics*. New York: Simon & Schuster.

Risjord, Mark. 2000. *Woodcutters and witchcraft: rationality and interpretive change in the social sciences*. Albany: State University of New York Press.

Rohwer, Yasha. 2014. "Lucky understanding without knowledge." *Synthese* 191 (5):945–959. doi: 10.1007/s11229-013-0322-0.

Rohwer, Yasha, and Collin C. Rice. 2013. "Hypothetical pattern idealization and explanatory models." *Philosophy of science* 80 (3):334–355. doi: 10.1086/671399.

Sainsbury, R. M. 1997. "Easy possibilities." *Philosophy and phenomenological research* 57 (4):907–919. doi: 10.2307/2953809.

Salmon, Wesley C. 1984. *Scientific explanation and the causal structure of the world.* Princeton: Princeton University Press.

Salmon, Wesley C. 1989. "Four decades of scientific explanation." In *Scientific Explanation*, edited by Philip Kitcher and Wesley C. Salmon, 3–219. Minneapolis: University of Minnesota Press.

Schoener, Thomas W., David A. Spiller, and Jonathan B. Losos. 2001. "Predators increase the risk of catastrophic extinction of prey populations." *Nature* 412 (6843):183–186.

Schrenk, Markus A. 2004. "Galileo vs. Aristotle on free falling bodies." *Logical analysis and the history of philosophy* 7.

Schurz, Gerhard, and Karel Lambert. 1994. "Outline of a theory of scientific understanding." *Synthese* 101 (1):65–120.

Scriven, Michael. 1959. "Explanation and prediction in evolutionary theory." *Science* 130 (3374):477–482.

Sellars, Wilfrid. 1963. *Science, perception and reality.* New York: Routledge & K. Paul.

Shadish, William R., Thomas D. Cook, and Donald Thomas Campbell. 2001. *Experimental and quasi-experimental designs for generalized causal inference.* Boston: Houghton Mifflin.

Skyrms, Brian. 1980. *Causal necessity: a pragmatic investigation of the necessity of laws.* New Haven: Yale University Press.

Sober, Elliot. 1986. "Explanatory presupposition." *Australasian journal of philosophy* 64:143–149.

Sosa, Ernest. 1980. "The raft and the pyramid: coherence versus foundations." *Midwest studies in philosophy* 5:3–25.

Sosa, Ernest. 1991. *Knowledge in perspective: selected essays in epistemology.* Cambridge: Cambridge University Press.

Sosa, Ernest. 1999. "How to defeat opposition to Moore." *Noûs* 33:141–153.

Sosa, Ernest. 2007. *Apt belief and reflective knowledge, volume 1: a virtue epistemology.* Oxford: Oxford University Press.

Sosa, Ernest. 2009. *Apt belief and reflective knowledge, volume 2: reflective knowledge.* Oxford: Oxford University Press.

Steel, Daniel. 2013. "Acceptance, values, and inductive risk." *Philosophy of science* 80 (5):818–828. doi: 10.1086/673936.

Strevens, Michael. 2000. "Do large probabilities explain better?" *Philosophy of science* 67 (3):366–390.

Strevens, Michael. 2008. *Depth: an account of scientific explanation.* Cambridge: Harvard University Press.

Strevens, Michael. 2013. "No understanding without explanation." *Studies in history and philosophy of science part A* 44 (3):510–515. doi: 10.1016/j.shpsa.2012.12.005.

Strevens, Michael. 2016. "How idealizations provide understanding." In *Explaining understanding: new essays in epistemology and philosophy of science*, edited by Stephen R. Grimm, Christoph Baumberger, and Sabine Ammon, 37–49. New York: Routledge.

Suárez, Mauricio. 2009. *Fictions in science: philosophical essays on modeling and idealization, Routledge studies in the philosophy of science*. New York: Routledge.

Sullivan, Emily. Manuscript. "Transforming knowledge into understanding." Available from the author.

Sullivan, Emily, and Kareem Khalifa. Manuscript. "Do idealizations provide non-factive understanding?" Available from the author.

Thagard, Paul. 1978. "The best explanation: criteria for theory choice." *Journal of philosophy* 75:76–92.

Thagard, Paul. 1992. *Conceptual revolutions*. Princeton: Princeton University Press.

Thagard, Paul. 1999. *How scientists explain disease*. Princeton: Princeton University Press.

Thagard, Paul. 2003. "Pathways to biomedical discovery." *Philosophy of science* 70 (2):235–254.

Thalos, M. 2002. "Explanation is a genus: an essay on the varieties of scientific explanation." *Synthese* 130 (3):317–354.

Trout, J. D. 2002. "Scientific explanation and the sense of understanding." *Philosophy of science* 69:212–233.

Trout, J. D. 2005. "Paying the price for a theory of explanation: de Regt's discussion of Trout." *Philosophy of science* 72:198–208.

Trout, J. D. 2007. "The psychology of scientific explanation." *Philosophy compass* 2/3:564–591.

Turner, Stephen P. 2013. "Where explanation ends: understanding as the place the spade turns in the social sciences." *Studies in history and philosophy of science part A* 44 (3):532–538. doi: 10.1016/j.shpsa.2012.12.001.

Turri, John. 2016. "Knowledge as achievement, more or less." In *Performance epistemology*, edited by Miguel A. Fernandez, 124–135. Oxford: Oxford University Press.

van Fraassen, Bas C. 1980. *The scientific image*. New York: Clarendon Press.

Vickers, Peter. 2013. *Understanding inconsistent science*. Oxford: Oxford University Press.

Wilkenfeld, Daniel A. 2013. "Understanding as representation manipulability." *Synthese* 190 (6):997–1016. doi: 10.1007/s11229-011-0055-x.

Wilkenfeld, Daniel A. 2014. "Functional explaining: a new approach to the philosophy of explanation." *Synthese* 191 (14):3367–3391. doi: 10.1007/s11229-014-0452-z.

Wilkenfeld, Daniel A., and Jennifer K. Hellmann. 2014. "Understanding beyond grasping propositions: A discussion of chess and fish." *Studies in history and philosophy of science part A* 48 (0):46–51. doi: 10.1016/j.shpsa.2014.09.003.

Wilkenfeld, Daniel A., Dillon Plunkett, and Tania Lombrozo. 2016. "Depth and deference: when and why we attribute understanding." *Philosophical studies* 173 (2):373–393. doi: 10.1007/s11098-015-0497-y.

Williamson, Timothy. 2000. *Knowledge and its limits*. Oxford: Oxford University Press.

Woodward, James. 2002. "Explanation." In *Blackwell guide to the philosophy of science*, edited by Peter Machamer and Michael Silberstein, 37–54. Malden: Blackwell.

Woodward, James. 2003. *Making things happen: a theory of causal explanation*. New York: Oxford University Press.

Woodward, James, and Christopher R. Hitchcock. 2003. "Explanatory generalizations, part I: a counterfactual account." *Noûs* 37 (1):1–24.

Worrall, John. 1989. "Structural realism: the best of both worlds?" *Dialectica* 43 (1–2):99–124.

Wray, K. Brad. 2001. "Collective belief and acceptance." *Synthese* 129 (3):319–333. doi: 10.1023/a:1013148515033.

Zagzebski, Linda. 1996. *Virtues of the mind: an inquiry into the nature of virtue and the ethical foundations of knowledge*. New York: Cambridge University Press.

Zagzebski, Linda. 2001. "Recovering understanding." In *Knowledge, truth, and duty: essays on epistemic justification, responsibility, and virtue*, edited by Matthias Steup, 235–252. Oxford: Oxford University Press.

Index

abilities, 13, 18, 20, 21, 51–79, 140, 185,
   226, 233
   Classic Ability Argument, 52, 53, 54, 61, 63, 64,
     67, 70, 73, 79
   Classic Ability Question, 18, 20, 51, 52, 53,
     60, 61
   cognitive, 63, 66, 190, 194, 216, 218, 219,
     225, 228
   inferential, 62, 63, 171, 190, 216
   involved in SEEing, 62–63
   linguistic, 56, 70, 144–50
   perceptual, 62, 190, 216
   special, 18, 52–57, 58, 59, 60, 61, 62, 63–65, 66,
     68, 74, 234
   Updated Ability Argument, 64, 69, 70
   Updated Ability Question, 20, 52, 53, 63, 64,
     65, 66
ability intuition, 62
acceleration, 135–39
acceptance, 13, 155, 169–73, 175–81, 182, 229
   and belief, 169–73, 175–81
achievements, 65, 68, 69, 216, 217, 218, 219, 226,
    227, 233, 234, *See also* abilities
   cognitive, 23, 65–70, 86, 92, 123, 129, 162, 165,
     166, 200, 216, 217, 218, 219, 225–28, 229, 234
   General Achievement Argument, 227, 231, 233
   how-why, 69
   Scientific Achievement Argument, 217, 218,
     226, 227, 231
   why-how, 67, 69, 70, 79
Achinstein, Peter, 24, 171
alpha-particles, 118
analogy, 25, 128, 140, 143, 152
   tacit, 140, 142, 143, 144
Anolis sagrei. See brown anole
antirealism. *See* realism and antirealism
Aristotle, 17, 136
Austere Breadth, 84, 100–4
   Austere Breadth Argument, 102
Austere Coherence, 85, 111–23, *See*
    Indeterministic Systems Argument

bacteria, 46, 141, 204–7
   gammaproteo-, 46
   Helicobacter pylori, 205, 207
Bahamas, 71, 72, 87–92, 100–4, 124
Bartelborth, Thomas, 142
Batterman, Robert, 27, 31, 32, 36, 42–43, 50
belief, 12, 13, 14, 25, 26, 42, 57, 58, 59, 60, 62, 65,
    66, 67, 68, 69, 74, 75, 76, 77, 88, 89, 90, 101,
    156, 159, 167, 169, 180, 202, 218, 225, 226,
    See also acceptance; luck; truth
Bernoulli's principle, 73
best available science, 12, 13, 22, 48, 89, 178, 234
Bjorken, James, 26, 27–41, 42, 43, 44, 50, 51
body of information. *See* subject matter
Bogen, Jim, 47
Bokulich, Alisa, 167
Boltzmann, Ludwig, 46, 167
Bonjour, Laurence, 73, 75, 76
Boyle's law, 47
Brandom, Robert, 59
Braverman, Mike, 180
Brogaard, Berit, 94
brown anole, 71, 72, 87–92, 101, 102, 103, 109,
    113, 124

Campbell, Donald, 203
Cao, Tian Yu, 34, 36
Carter, J. Adam, 18, 65
   on epistemic value, 212, 226–28, 229, 231, 233
   on objectual understanding, 81, 84, 92–100,
    101, 105, 109, 110, 112
Cartwright, Nancy, 8, 108
Cavendish, Henry, 160, 163, 165
Chang, Hasok, 159–64
Chisholm, Roderick, 186
Code, Lorraine, 62
cognitive benefits, 126, 127, 128, 132, 135, 137, 138,
    139, 143, 147, 151, 152, 165, 168
Cohen, L. Jonathan, 170, 171, 180
combustion, 160, 166
concept-possession, 57, 58, 59, 74, 78

confounds, 200, 201–4, 205, 211
conservatism, 25, 160
Cook, Thomas, 203
counterfactual reasoning, 58, 59, 60, 73–75, 76, 78, 79
Craver, Carl, 47, 97, 118
critical information, 135–39
cross-section (σ), 27, 28, 29
 σ_{DIS}, 29, 34, 37, 40
 σ_{MOTT}, 29, 37
current algebra, 28, 36, 40, 45
Cushing, James, 116

Darden, Lindley, 97
David, Marian, 215
De Regt, Henk, 17, 18, 19, 27, 42, 43–50, 52, 65, 71, 154, 158–62, 163, 165, 167
deflationism about understanding, 21, 22, 234
DePaul, Michael, 183
dependency relations, 72, 73
Devitt, Michael, 76
Diéguez, Antonio, 167
Dieks, Dennis, 17, 47, 48, 52, 65
Díez, José, 24
difference-fakers, 174, 175, 178, 179
domain. *See* subject matter
Douglas, Heather, 109, 171
Doyle, Yannick, 154, 167, 168

Egan, Spencer, 154, 167, 168
EKS (Explanation-Knowledge-Science) Model, 14–15, 20–21, 22, 23–25, 26, 27, 38, 40–45, 48–50, 51, 52, 54, 61, 63–68, 69, 70–75, 76, 79, 81, 82, 87, 96, 102, 103, 104, 106, 107, 110, 116, 123, 124, 126, 130, 131, 136, 137, 139, 144, 149, 150–53, 154, 155, 156, 158, 164, 165, 166, 169, 170, 173, 176, 177, 181, 184, 186, 188, 189, 195, 208, 209, 210, 213, 214, 218, 222, 226, 228
 (EKS1), 14, 15, 23, 53, 84, 91, 95, 126, 150
 (EKS2), 14, 15, 21, 24, 52, 55, 56, 57, 58, 81, 82, 84, 101, 103, 106, 126, 130, 155, 156, 176, 177, 215, 221
ceteris paribus clause, 4, 15, 64
electrons, 27–30, 32, 33, 34, 37, 38, 39, 41, 113–15, 116, 118, 119–22, 123, 163, 164
Elgin, Catherine, 18, 19, 81, 84, 154, 158, 167, 172
empirical adequacy, 7, 24, 25, 38, 44
empiricism, 2, 76, 79, 161
enablers. *See* abilities
epistemic hoarding, 98, 99, 101, 110, 121, 124
epistemic status, 14, 60, 212, 223, 232, 233
epistemic value, 19, 21, 45, 50, 93, 212–35, *See also* achievements; explanation, value of
 Basic Value Question, 213–14, 218, 219–21, 222
 distinctive, 213, 216, 221–33, 234
 Distinctive Value Question, 221–25, 226–33
 final, 215, 216, 217, 218, 219, 224, 225, 227, 230–33
 Final Value Motivation, 224, 229, 231, 232
 first-order evaluations, 214
 fundamental, 215, 218, 219, 234
 General Value Question, 19, 21, 213, 214, 219–21, 233
 instrumental, 44, 215, 216, 218, 219, 226, 234
 second-order evaluations, 214
European Center for Particle Physics (CERN), 41
evidence, 4, 11, 12, 13, 14, 22, 25, 26, 35, 38, 40, 45, 49, 62, 69, 89, 91, 97, 102, 104, 109, 139, 173, 178, 180, 193, 194–201, 206, 209, 216, 218, 227
 and experimental design, 201–4
exothermic reactions, 93, 94, 110, 111
experiment, 3, 22, 23, 26, 27, 28, 29, 30, 33, 34, 35, 38, 40–43, 45, 49, 50, 63, 64, 68, 72, 76, 90, 160, 165, 172, 201, 203, 204–7, 208
explanation
 "theory" of, 6–8, 21, 55, 108, 116, 156, 163
 asymptotic, 8, 24, 31–34, 35, 36, 39, 42, 43, 44, 116
 causal, 8, 16, 17, 24, 47, 63, 76, 88, 107, 108, 112, 115, 117, 118, 123, 127, 132, 145, 146, 172, 202
 contrastive, 8, 89, 109, 115, 117, 119–22, 123
 correct, 6, 9, 10, 12, 14, 24, 31, 42, 44, 45, 54, 56, 57, 65, 68, 71, 87, 88, 109, 113, 125, 127, 128, 129–35, 136, 137, 150–53, 159, 165, 166, 189, 214, 218, 225, 226, 230
 deductive-nomological, 17, 47, 107
 deductive-nomological probabilistic (DNP), 108, 117, 120
 difference-making, 7, 38–40, 54–57, 59, 120, 122, 140, 145, 156–58, 163, 164, 166, 168, 169, 173, 174, 175, 176, 178
 explanation façades. *See* confounds
 Explanation Question, 19, 21, 125, 153
 heavyweight theories of, 21
 indeterministic. *See* explanation, probabilistic
 inter-explanatory relationships, 6, 9–10, 63, 90
 local constraints, 7, 8, 21, 38, 55, 108, 141, 142, 145, 157
 mechanistic, 8, 16, 17, 25, 47, 55, 69, 88, 97
 non-causal, 8, 116
 ontological requirements, 7, 38, 55, 141, 157, 158, 163, 173, 175
 pluralism, 8, 24, 108, 215
 potential, 12, 13, 25, 26, 27, 40, 45, 62, 75, 89, 90, 91, 109, 128, 131–38, 152, 161, 173, 178, 179, 197, 199
 probabilistic, 8, 108, 109, 110, 111, 113, 114, 115, 116–19

explanation (cont.)
 schema, 31, 32, 35, 36, 37, 38, 44, 118, 120,
  140–44
 value of, 215, 219, 221
explanationism, 85, 86, 115, 123
explanatory nexus, 6, 9, 11, 12, 14, 15, 24, 51, 64, 68,
  71, 81, 87, 107, 134, 138, 150, 151, 177, 184, 213,
  214, 216
Explanatory Objection, 130, 140, 142, 143, 145,
  147, 149, 150
 and quasi-factive understanding, 161, 162–64,
  166, 168, 177, 180, 181
explanatory roles, 86–92, 96, 98, 99, 100, 102, 104,
  107–11, 114, 121, 129, 133, 135–38, 139, 143, 147,
  165, 166, 168
 direct, 87, 88, 89, 90, 96, 108, 133, 134, 147
 indirect, 86–92, 96, 97, 102, 109, 110, 114, 133,
  134, 137
explanatory understanding, 2, 3, 5, 6, 16, 18, 20,
  53, 56, 65, 70, 125, 129, 130, 132, 134, 139, 143,
  146, 147, 150–53, 158–62, 164–66, 168, 180,
  195, 215, 219, 227, 228
 versus objectual understanding, 80–124
externalism, 57

fair comparisons, 94, 101, 106, 110, 114, 123
fatalism, 223, 224, 225–26, 230, 233, 234
 hyper-, 224, 225, 229
 therapeutic, 225, 229
Faye, Jan, 171
Feynman, Richard, 27, 35–41, 43, 45, 50, 51
 impulse approximation, 37, 38
 infinite momentum frame, 36, 38
Friedman, Michael, 108, 141, 142
fruitfulness, 25, 62

Gadomski, Michael, 183
Galilei, Galileo, 135–39
Garfinkel, Alan, 88, 109, 171
Gendler, Tamar Szabó, 137
Gijsbers, Victor, 18, 82
Ginet, Carl, 186
Glymour, Bruce, 117, 120, 121
Goldman, Alvin, 186, 197
Gordon, Emma, 18, 65
 on epistemic value, 212, 226–28, 229, 231, 233
 on objectual understanding, 81, 84, 92–100,
  101, 105, 109, 110, 112
Graham, Noah, 154, 167, 168
grasping, 3, 6, 9, 10, 11, 14, 24, 31, 51, 64, 68, 73, 75,
  77, 79, 81, 84, 85, 87, 91, 95, 98, 101, 104, 105,
  106, 107, 108, 110, 111, 112, 113, 114, 121, 125,
  126, 128, 129, 130, 131, 133, 136, 137, 143, 145,
  150, 165, 177, 184, 213, 214, 216
 completeness of, 10, 13

Greco, John, 13, 17, 62, 73, 154, 167, 183, 185, 216
Grimm, Stephen, 16, 17, 24, 52, 58, 65, 72–79, 154,
  183, 215, 233
 modal apprehension, 73, 75–79

Hacking, Ian, 8
hadrons, 26, 27, 33, 35, 36, 37, 38, 41
Harman, Gilbert, 22
Hazen, T. C., 46
Hellmann, Jennifer, 52
Hempel, Carl, 16, 17, 21, 47, 107, 108, 116, 204
Hills, Alison, 18, 52, 58, 65, 70–72, 79, 183
 cognitive control, 70, 79
Hindriks, Frank, 109, 125, 167
Historical Argument, 155, 161, 166, 180, 181
Hitchcock, Christopher, 88, 108, 109, 117,
  119–22, 204
Humphreys, Paul, 117
Hurricane Floyd, 87, 88, 90, 91, 101, 102, 104

ideal gas law, 47, 166–81
idealization, 8, 55, 154, 155, 166–81, 229
 idealized explanation, 167, 168, 169, 174, 175,
  176, 180
improbable explananda, problem of, 46–48, 71
improvements to understanding, 213, 214,
  219, 224
 explanatory improvement, 214, 215, 218, 225,
  226, 229
 improved comparison, 216, 219, 225, 226
 improved consideration, 216, 219, 225, 226, 229
 scientific improvement, 216, 217, 218, 225, 226,
  229, 234
Indeterministic System Argument, 113–15, 123
inference to the best explanation, 12, 22
internalism, 233
intervention, 8, 172, 201, 202, 203, 204
 ideal, 201, 202, 203, 211
invariance, 204
irrelevant insights, problem of, 45–46

Jeffrey, Richard, 108
justification, 56, 75, 77, 185, 204, 233

Kelp, Christoph, 4, 18, 82, 84, 101, 183
Kepler's laws, 108
Khalifa, Kareem, 24, 44, 65, 104, 154, 167,
  168, 183
Kim, Jaegwon, 72, 73
Kitcher, Philip, 17, 24, 108, 117, 140–42, 204
knowledge. *See* scientific knowledge
 a priori, 75, 76, 77, 139
 causal, 67, 144, 145, 147, 148
 de dicto, 77, 78, 79
 de re, 77, 78

explanatory, 16, 18, 21, 24, 25, 42, 52, 53, 54, 56, 62, 63, 66, 73, 78, 80, 82, 92, 104, 109, 133, 134, 135, 140, 151, 153, 155, 190, 195, 212, 219
inferential, 62, 143, 216
modal, 128, 131, 135–39
non-propositional, 74, 76, 77, 78
of explananda, 29–30
semantic, 145, 147–50
tacit, 128, 140, 142, 143–45, 147, 148, 149
knowledge-how, 51, 70, 77, 78, 79
knowledge-why. *See* knowledge, explanatory
Kuhn, Thomas, 140, 143, 144
Kvanvig, Jonathan, 18, 19, 154, 158, 167, 179, 180, 183, 187, 212, 213, 223, 233
on objectual understanding, 81, 82, 84, 85, 111–23

Lackey, Jennifer, 219
Lambert, Karel, 141
Laudan, Larry, 159
Leiocephalus carinatus. See northern curly-tailed lizard
leptons, 27, 33, 38
Leuridan, Bert, 24
Lewis, David, 7, 117
Lipton, Peter, 12, 16, 19, 22, 24, 42, 108, 125–53, 203, 204
Lipton's Argument, 126–31, 135, 138
Lipton's Assumption, 126–31, 134, 135, 138, 140, 143, 144, 147, 150–53
Lombrozo, Tania, 11
Losos, Jonathan, 71, 87, 89, 90
luck, 19, 20, 21, 22, 56, 65, 69, 178, 183–211, 233, 234, *See also* luck compatibilism; luck incompatibilism; safety
and belief, 184–201, 207–11
Barn Façade Example, 186–88, 189, 190, 191, 192, 193, 196, 197
benign, 193, 194
Classic Luck Question, 19, 20, 183, 184, 185, 186, 189
environmental, 186–88, 190, 191, 193, 194, 197, 209
Environmentally Lucky Understanding Argument, 188, 189
evidential, 193, 194, 197
Fiona and Carmen Example, 196, 198
Gettier, 12, 19, 178, 183, 186–87, 190, 209–11, 212
Gettier-Lucky Understanding Argument, 210
Jesse James Example, 193
Nellie Example, 210
Nero Example, 187–88, 189–94, 196
Sheep Example, 186, 210
Strict Nero Example, 191, 192, 194
Updated Luck Question, 20, 184, 209

luck compatibilism, 194, 195, 208
classic, 185, 186, 188, 189, 194, 208, 210
updated, 185, 208, 209
luck incompatibilism, 187, 188, 194
classic, 185, 188, 189
scientific, 195, 200, 208, 209, 217
updated, 185, 186, 188, 189, 194, 195, 208
Lycan, William, 22, 108, 203
Lynch, Michael, 215

Machamer, Peter, 97
manipulation, 144, 148, 152, 160, 172
Marshall, Barry, 204–7
memory, 18, 53, 56, 62, 63, 73, 77, 190
Meno, 212
Millson, Jared, 24
Mitchell, Sandra, 203
Mizrahi, Moti, 154, 158, 167
Morris, Kevin, 56, 183, 209–11

naturalism, 76, 79
neutrons, 27
Newman, Mark, 18, 52, 56, 65
Newtonian mechanics, 108
Nexus Principle, 6–10, 13, 14, 24, 31, 34–35, 38, 45, 46, 48, 50, 68, 87, 96, 97, 107, 177, 214
non-factivism, 154, 155, 156–66, 167, 172, 173–75, 176, 177, 180, 181–82
northern curly-tailed lizard, 71, 87–92, 102
Norton, John, 116
Nozick, Robert, 185, 193

objectual understanding, 3, 5, 18, 20, 80–124, 154, 227, 228, 233, 234, *See also* explanationism; proto-understanding; quasi-explanationism; objectualism
breadth. *See* Austere Breadth; Robust Breadth
Classic Objectual Question, 18, 20, 80, 81, 83, 85, 93, 106
coherence. *See* Austere Coherence; Robust Coherence
minimal, 101, 102, 104
Updated Objectual Question, 20, 81, 82, 84, 92, 95, 103
objectualism, 104
austere, 82, 83, 84, 86, 96, 100–4, 111–15, 123
classic, 83
classic austere (CAO), 83
classic robust (CRO), 83, 93, 105, 106
Kvanvig's, 112
robust, 20, 81, 83, 84, 86, 92–100, 101, 104–11, 112
updated, 85
updated austere (UAO), 84
updated robust (URO), 84

Orozco, Joshue, 185
outSEEing, 199, 200, 211
overdetermination, 132
oxygen, 9, 53, 56, 57, 58, 59, 60, 66, 67, 68, 74, 75, 77, 78, 160, 161, 163, 164, 165

Parroting Objection, 57, 59
partons, 27, 30, 35–42, 43, 49
Paschos, Emmanuel, 34, 35
Peacocke, Christopher, 139
pessimistic induction, 159
p-façades, 191, 192
Philosophical Focus Motivation, 224, 225, 229, 231, 235
phlogiston, 158–66
photons, 30, 37, 40, 122, 123
Pickering, Andrew, 34
Plunkett, Dillon, 11
Popper, Karl, 204
possible worlds, 26, 184, 190, 191, 192, 194, 197, 198, 199, 206, 211
prediction, 26, 27, 36, 63, 89, 109, 159, 160, 200
Priestley, Joseph, 159, 160, 163
Pritchard, Duncan, 12, 18, 19, 26, 52–54, 56, 58, 62, 65–70, 72, 79, 105, 154, 158, 172, 178, 184, 186, 198, 210, 219
    epistemic grip, 53, 66, 67, 68, 69
    on epistemic value, 212, 217, 218, 221–25, 226–28, 229, 233
    on lucky understanding, 183, 185, 186, 187–94, 196, 210
protons, 27–30, 33, 36, 37, 38, 40, 41, 80
proto-understanding, 85–92, 100–4, 113, 115, 124, 129, 132, 133, 134, 136, 139, 143, 147, 150, 161, 165, 166
Putnam, Hilary, 164

quantum, 3, 20, 28, 80, 111, 112, 113, 114, 116, 118, 119, 122, 138
    electrodynamics (QED), 28, 29, 30, 35, 40
quarks, 35, 36, 37, 38
quasi-explanationism, 85–87, 92, 95–100, 103, 104, 108, 110, 111, 113, 114, 115, 120, 123, 124, 228
quasi-factivism, 154, 155, 156, 158, 162, 163, 167, 172, 180, 181, *See also* Explanatory Objection; Right Track Objection; Splitting Strategy; Swelling Strategy; Wrong Benefit Objection

Railton, Peter, 108, 117, 118, 119, 120
realism and antirealism, 7, 8, 12, 24, 25, 38, 55, 157, 158, 162, 163, 164, 168, 173, 215
received view of understanding, 16–22, 24, 50, 51–52, 53, 54, 61, 64, 66, 68, 73–75, 78, 79,

80–82, 89, 107, 124, 125, 126, 154–55, 156, 164, 176, 183–86, 188, 189, 194, 209, 213, 234
reference, 163, 164, 166, 168, 173, 176
Regge exchange, 41, 43, 45, 49
retrograde motion, 144, 149
revisionism, 223, 232, 233
Riaz, Amber, 183
Rice, Collin, 39, 167
Riggs, Wayne, 19, 154
Right Track Objection, 129–38, 143–45, 147–50, 151
    and quasi-factive understanding, 161, 165, 166, 168, 177, 180, 181
Riordan, Michael, 34, 36, 45
Risjord, Mark, 24, 88, 109, 171
Robust Breadth, 84, 92–100, 106
    Classic Breadth Argument, 93, 95, 106
    Robust Breadth Argument, 96, 97
Robust Coherence, 84, 104–11
    Classic Coherence Argument, 105, 106
    Robust Coherence Argument, 107, 110
Rohwer, Yasha, 167, 183, 210
rust, 160, 162, 164, 165

safety, 12, 13, 25, 26, 48, 49, 50, 89, 173, 178, 179, 180, 184, 185, 188, 189, 194, 195, 196, 200, 210, 218, 219, 225, 226
    comparative, 185, 189, 194–201, 203, 207–11, 216, 218, 234
    in experimental design, 201–7
    Safety Principle, 184, 190, 198
    Scientific Safety Argument, 195, 200
Sainsbury, Mark, 12, 178, 184
Sakurai, J. J., 41, 43
Salmon, Wesley, 16, 17, 21, 24, 107, 108, 117, 191
scaling, 26, 29–30, 31, 32, 34, 35, 36, 37, 38, 40–42
    Bjorken, 29, 30, 32, 33, 34
scattering, 26, 27–29, 31, 33, 34–42, 80
    elastic, 28
    inelastic, 28, 30, 32, 34, 42
Schoener, Thomas, 71, 87, 89, 90
Schrenk, Markus, 137
Schurz, Gerhard, 141
scientific explanatory evaluation, 12, 13, 25, 26, 48, 62–63, 68, 69, 71, 73, 74, 75, 76, 89, 110, 178, 179, 216, 218, 219, 225, 226, *See also* outSEEing
    and safety, 194–209
    belief-formation, 12, 13, 25, 40, 62, 63, 89, 179, 196, 197, 202, 206
    comparison of explanations, 12, 13, 22, 25, 26, 40, 62, 63, 71, 75, 89, 90, 136, 137, 178, 196, 197, 199, 202, 207, 216, 220, 225, 234

consideration of explanations, 12, 13, 25, 26, 40,
  62, 63, 71, 75, 89, 90, 136, 178, 196, 197, 202,
  216, 225
  Galileo's, 135–38
  in particle physics, 40–43
scientific goals, 14, 170, 171, 172, 173, 175, 178,
  179, 229
scientific knowledge, 9, 11–13, 14, 15, 18, 20, 21, 25,
  26, 29, 35, 40, 45, 52, 54, 61–65, 68, 70, 79,
  87, 89, 92, 95, 97, 103, 124, 126, 129, 133, 137,
  138, 147, 151, 165, 169, 170, 178, 195, 209, 216,
  230
  resemblance to, 11, 13–14, 15, 21, 24, 51, 64, 81,
    87, 149, 150, 173, 177, 184, 213, 216
  with acceptance instead of belief, 175–81
Scientific Knowledge Principle, 10–14, 24, 35,
  48–50, 64, 68, 69, 87, 96, 109, 177, 195, 216
scope, 25, 62, 203
Scriven, Michael, 47, 71
SEEing. *See* scientific explanatory evaluation
Sellars, Wilfrid, 11, 59
Shadish, William, 203
simplicity, 12, 25, 62
Skyrms, Brian, 203
Sober, Elliot, 88, 109
Socrates, 212
Sosa, Ernest, 12, 13, 62, 178, 184, 185, 216
Spiller, David, 71, 87, 89, 90
Splitting Strategy, 169, 170, 173–75, 177, 180, 181
Stanford Linear Accelerator (SLAC), 27, 28,
  30, 41
  SLAC-MIT group, 27, 28, 29, 30, 34, 38
statistical mechanics, 167, 174
statistical reasoning, 76, 206, 207
Steel, Daniel, 180
stoichiometry, 94, 110, 111
Strevens, Michael, 7, 121, 154, 167, 173
Suárez, Mauricio, 154
subject matter, 3, 18, 80, 82–85, 86, 92–95, 96,
  100, 101, 105, 110, 172, 227, 228
Sullivan, Emily, 57, 76, 167, 230
Swelling Strategy, 169, 170, 175–81, 182
syphilis-paresis example, 47, 48, 71

tacit explanations, 70, 140, 142, 145–47, *See*
  knowledge, tacit
Taylor, Richard, 41
testability, 25
testimony, 18, 20, 53, 54, 56, 60, 63, 73, 77, 192,
  193, 209
Thagard, Paul, 22, 31, 108, 141, 203, 204
Thalos, Mariam, 116
theoretical virtues, 25
thought experiments, 92, 102, 135–39, 194, 201
Trout, J. D., 17

truth, 154–82, 233, 234, *See also* explanation,
  ontological requirements; Historical
  Argument; idealization; non-factivism;
  quasi-factivism; realism and antirealism
  and belief, 12, 19, 25, 30, 55, 60, 63, 65–70, 79,
    88, 89, 100–4, 106, 155, 183, 185, 186, 187, 190,
    195, 198, 207–9, 210, 212, 214–19, 221,
    225–26, 228, 234
  approximate, 8, 12, 14, 21, 25, 38, 52, 54–57, 58,
    60, 81, 101, 103, 126, 127, 130, 141, 155, 156–58,
    159, 162, 163, 166–69, 173–75, 176, 189, 221
  Truth Question, 19, 21, 155
Turner, Stephen, 125
Turri, John, 218

ulcers, 204–7
understanding
  and context, 4, 5, 14, 15, 24, 35, 43, 46, 47, 48,
    51, 61, 79, 108, 116, 171, 188, 189, 196, 210,
    220, 221, 229
  better. *See* understanding, degrees of
  broad linguistic, 2
  comparative. *See* understanding, degrees of
  conceptual, 139
  degrees of, 3–5, 6, 10, 11, 13–14, 15, 20, 25, 35, 52,
    54, 56, 59, 61, 62, 63, 64, 81, 83, 84, 91, 95, 96,
    126, 150, 151, 177, 179, 184, 185, 189, 194, 195,
    200, 207–9, 213, 214, 218, 221, 228–30, 231
  everyday, 15, 195, 196
  explanatory. *See* explanatory understanding
  generic, 5, 15, 180, 188
  ideal, 3, 4, 5, 15, 35, 61, 84, 151, 153, 188, 214, 220,
    231, 232
  increase in. *See* improvements to
    understanding
  kinds of, 2–3
  minimal, 3, 4, 5, 6, 14, 15, 21, 24, 35, 51, 52,
    54–60, 61, 65, 72, 74, 81, 82–85, 100–4, 106,
    113, 126, 130, 155, 156, 164, 170, 176, 177, 188,
    189, 215, 219, 220, 221
  narrow linguistic, 2
  nature of, 19, 183, 212, 228, 233
  non-explanatory interrogative, 2, 82
  objectual. *See* objectual understanding
  of empirical phenomena, 2, 3, 11, 76, 82, 107,
    126, 139, 195
  outright, 4, 5, 15, 34, 35, 60, 61, 188, 208, 210,
    219–21, 227, 231, 233
  procedural, 2, 143, 151, 152, 165
  propositional, 2, 82, 112
  qualitative, 44–45
  received view. *See* received view of
    understanding
  understanding-why. *See* explanatory
    understanding

understanding (cont.)
  value of. *See* epistemic value
  without explanation. *See* understanding
    without explanation
understanding without explanation, 125–53, 234
  UWE, 125, 126, 128, 131, 134, 138, 139, 150
Unger, Peter, 186
unification, 8, 16, 17, 24, 25, 55, 108, 127, 128, 130,
    143–44, 161
  analogical, 140, 141, 143, 144

validationism, 223, 230, 232, 233
  objections to, 226–28
value of understanding. *See* epistemic value
van Fraassen, Bas, 7, 24, 25, 44, 47, 171, 215
vector mesons, 41, 45, 49
Vickers, Peter, 137
virtue epistemology, 13, 22, 62, 216, 217, 219
visualization, 45, 116, 128, 144, 145, 152

Warren, Robin, 204–7
What-If Objection, 58, 59
what-if questions, 58, 59, 60, 74, 89, 97, 133,
    134, 200
what-if reasoning. *See* counterfactual
    reasoning
Wilkenfeld, Daniel, 11, 17, 52, 56
Williamson, Timothy, 12, 178, 184
Woodward, James, 7, 24, 58, 89, 108, 116–22, 133,
    142, 172, 201–4, 206
Worrall, John, 8
Wray, K. Brad, 180
Wrong Benefit Objection, 130, 135, 138, 139,
    143, 150
  and quasi-factive understanding, 161, 164–66,
    168, 177, 180, 181

Zagzebski, Linda, 13, 62, 154, 167, 183,
    212, 216